Some Recent Advances in Partial Difference Equations

Edited by: Eugenia N. Petropoulou

Bentham Science Publishers, 2010

eBooks End User License Agreement

Please read this license agreement carefully before using this eBook. Your use of this eBook/chapter constitutes your agreement to the terms and conditions set forth in this License Agreement. Bentham Science Publishers agrees to grant the user of this eBook/chapter, a non-exclusive, nontransferable license to download and use this eBook/chapter under the following terms and conditions:

1. This eBook/chapter may be downloaded and used by one user on one computer. The user may make one back-up copy of this publication to avoid losing it. The user may not give copies of this publication to others, or make it available for others to copy or download. For a multi-user license contact permission@bentham.org

2. All rights reserved: All content in this publication is copyrighted and Bentham Science Publishers own the copyright. You may not copy, reproduce, modify, remove, delete, augment, add to, publish, transmit, sell, resell, create derivative works from, or in any way exploit any of this publication's content, in any form by any means, in whole or in part, without the prior written permission from Bentham Science Publishers.

3. The user may print one or more copies/pages of this eBook/chapter for their personal use. The user may not print pages from this eBook/chapter or the entire printed eBook/chapter for general distribution, for promotion, for creating new works, or for resale. Specific permission must be obtained from the publisher for such requirements. Requests must be sent to the permissions department at E-mail: permission@bentham.org

4. The unauthorized use or distribution of copyrighted or other proprietary content is illegal and could subject the purchaser to substantial money damages. The purchaser will be liable for any damage resulting from misuse of this publication or any violation of this License Agreement, including any infringement of copyrights or proprietary rights.

Warranty Disclaimer: The publisher does not guarantee that the information in this publication is error-free, or warrants that it will meet the users' requirements or that the operation of the publication will be uninterrupted or error-free. This publication is provided "as is" without warranty of any kind, either express or implied or statutory, including, without limitation, implied warranties of merchantability and fitness for a particular purpose. The entire risk as to the results and performance of this publication is assumed by the user. In no event will the publisher be liable for any damages, including, without limitation, incidental and consequential damages and damages for lost data or profits arising out of the use or inability to use the publication. The entire liability of the publisher shall be limited to the amount actually paid by the user for the eBook or eBook license agreement.

Limitation of Liability: Under no circumstances shall Bentham Science Publishers, its staff, editors and authors, be liable for any special or consequential damages that result from the use of, or the inability to use, the materials in this site.

eBook Product Disclaimer: No responsibility is assumed by Bentham Science Publishers, its staff or members of the editorial board for any injury and/or damage to persons or property as a matter of products liability, negligence or otherwise, or from any use or operation of any methods, products instruction, advertisements or ideas contained in the publication purchased or read by the user(s). Any dispute will be governed exclusively by the laws of the U.A.E. and will be settled exclusively by the competent Court at the city of Dubai, U.A.E.

You (the user) acknowledge that you have read this Agreement, and agree to be bound by its terms and conditions.

Permission for Use of Material and Reproduction

Photocopying Information for Users Outside the USA: Bentham Science Publishers Ltd. grants authorization for individuals to photocopy copyright material for private research use, on the sole basis that requests for such use are referred directly to the requestor's local Reproduction Rights Organization (RRO). The copyright fee is US $25.00 per copy per article exclusive of any charge or fee levied. In order to contact your local RRO, please contact the International Federation of Reproduction Rights Organisations (IFRRO), Rue du Prince Royal 87, B-I050 Brussels, Belgium; Tel: +32 2 551 08 99; Fax: +32 2 551 08 95; E-mail: secretariat@ifrro.org; url: www.ifrro.org This authorization does not extend to any other kind of copying by any means, in any form, and for any purpose other than private research use.

Photocopying Information for Users in the USA: Authorization to photocopy items for internal or personal use, or the internal or personal use of specific clients, is granted by Bentham Science Publishers Ltd. for libraries and other users registered with the Copyright Clearance Center (CCC) Transactional Reporting Services, provided that the appropriate fee of US $25.00 per copy per chapter is paid directly to Copyright Clearance Center, 222 Rosewood Drive, Danvers MA 01923, USA. Refer also to www.copyright.com

CONTENT

DEDICATION

In memory of Professor Panayiotis D. Siafarikas

Foreword

While "Mathematical Reviews" currently lists 1058 books containing "Partial Differential Equations" in their title and 128 books containing "Difference Equations" in their title, it only lists 3 books containing "Partial Difference Equations" in their title. On the other hand, 238 journal publications are listed containing "Partial Difference Equations" in their title, so research in this area is rather active and ongoing. This is due to the rich possibility of theoretical investigations and the numerous applications which partial difference equations enjoy. These facts illustrate that there is an urgent need to expand the availability of textbooks in the area of "Partial Difference Equations".

The book at hand, "Some Recent Advances in Partial Difference Equations", as edited and presented by Professor Eugenia Petropoulou, is a welcome, timely, and excellent contribution filling the above described gap. Professor Petropoulou has done a terrific job in putting together this volume, offering four chapters on distinct topics of current interest in the area of partial difference equations.

The first chapter covers oscillation theory of partial difference equations and is written by Professor Patricia Wong (Singapore), a world-wide leading expert in the area of differential, difference, and dynamic equations and in particular oscillation theory for these equations. Criteria for the nonexistence of positive solutions of certain partial difference equations with deviating arguments are presented and several examples are offered.

The second chapter shows a connection between functional analysis and partial difference equations and is written by the late Professor Panayiotis Siafarikas (Greece), the internationally esteemed expert in this area of research, together with his student Professor Eugenia Petropoulou (Greece), who is the editor of this volume. A functional-analytic method to study partial difference equations is developed and illustrated by two fundamental examples.

The third chapter discusses the connection of partial difference equations to systems theory and is written by Professor Jiří Gregor and Professor Josef Hekrdla (Czech Republic). Existence and uniqueness results for initial value problems and boundary value problems involving linear partial difference equations are presented and extended to systems of linear partial difference equations. These results are applied to input–output relations of linear multidimensional systems.

The fourth chapter offers some numerical schemes constituting partial difference equations and is written by Professor Efstratios Tzirtzilakis and Professor Nikolaos Kafoussias (Greece). Partial differential equations are discretized in order to obtain numerical schemes resulting in partial difference equations, and the connection of the solutions of these two equations is examined analytically and numerically.

Of course these covered topics only scratch the surface of this exciting area of research. We look forward to future developments inspired by the publication of this volume.

Martin Bohner, Rolla, Missouri (USA)
October 20, 2010

Preface

Partial difference equations arise naturally in many areas of science and are used for the description of many realistic problems, such as probability problems, problems in queuing theory, physical problems, biological problems, etc. It could be stated that partial difference equations are on one side of a coin, where on the other side stand partial differential equations. However, the main application of partial difference equations is probably in numerical analysis, where they arise naturally when discretizing a partial differential equation.

Lately, there is an increasing interest in partial difference equations demonstrated by the enormous amount of research papers devoted to them. The initial reason for this increasing interest was probably the development of computers and the area of numerical analysis. However, there are very few books devoted exclusively to partial difference equations in contrast to their continuous analog of partial differential equations on which hundreds of books have been written. This e-book is an attempt to present a few recent advances in partial difference equations including oscillation results, which are in the front line of research, functional analytic methods for studying partial difference equations, which are not so commonly used, as well as applications of partial difference equations in systems theory and numerical analysis. Basic techniques for solving or studying partial difference equations were not included, since these can be found in several chapters of books on difference equations. Of course, several other types of results could have been included, but this was not possible due to the lack of time. Probably in the future a more complete book on the subject, will be published!

The e-book constitutes of four chapters. Each chapter could be characterized as a review paper. Moreover, in all chapters there are several examples in order to help the interested reader get more acquainted with the presented methods and results.

I would like to express also from this position, my sincerest thanks to the colleagues who contributed in this e–book and made possible its publication. Also, I would like to thank the personnel of the Bentham Science Publishers and especially Bushra Siddiqui for their continuous collaboration and support.

Eugenia N. Petropoulou

Assist. Professor

CHAPTER 1

Oscillation of partial difference equations with deviating arguments

Patricia J. Y. Wong

School of Electrical and Electronic Engineering, Nanyang Technological University, 50 Nanyang Avenue, Singapore 639798, Singapore.
e-mail: ejywong@ntu.edu.sg

Abstract: We present criteria for the nonexistence of eventually positive (negative) and nondecreasing (nonincreasing) solutions of the partial difference equations

$$\nabla_m \nabla_n y(m,n) + P(m,n,y(m+k,n+\ell)) = Q(m,n,y(m+k,n+\ell))$$

and

$$\nabla_m \nabla_n y(m,n) + \sum_{i=1}^{\tau} P_i(m,n,y(m+k_i,n+\ell_i)) = \sum_{i=1}^{\tau} Q_i(m,n,y(m+k_i,n+\ell_i)).$$

Criteria are also established for the nonexistence of eventually positive (negative) and monotonely decreasing (increasing) solutions of the partial difference inequalities

$$\Delta_m \Delta_n y(m,n) + \sum_{i=1}^{\tau} P_i(m,n,y(g_i(m),h_i(n))) \geq (\leq) \sum_{i=1}^{\tau} Q_i(m,n,y(g_i(m),h_i(n)))$$

where $g_i(m)$ and $h_i(n)$, $1 \leq i \leq \tau$ are some deviating arguments. Several examples are included to dwell upon the importance of the results obtained.

Key words and phrases: Eventually positive solutions, monotone solutions, partial difference equations, deviating arguments.

Eugenia N. Petropoulou (Ed)

1.1   Introduction

The theory of difference equations, the methods used in their solutions, and their wide applications have been and still are drawing intense attention. In fact, in the last two decades several monographs and hundreds of research papers have been written, e.g., see [1–21] and the references cited therein. On the other hand, though partial difference equations are as important as difference equations, comparatively few papers have been devoted to the qualitative theory of their solutions, for instance, refer to [22–38]. In fact, partial difference equations arise in applications involving population dynamics with spatial migrations, chemical reactions, control systems, combinatorics as well as finite difference schemes [28], [39]. Hence, to further develop the qualitative theory of partial difference equations, in this paper we shall consider the partial difference equations

$$\nabla_m \nabla_n y(m, n) + P(m, n, y(m + k, n + \ell)) = Q(m, n, y(m + k, n + \ell)), \quad (1.1.1)$$

$$\nabla_m \nabla_n y(m, n) + \sum_{i=1}^{\tau} P_i(m, n, y(m + k_i, n + \ell_i)) = \sum_{i=1}^{\tau} Q_i(m, n, y(m + k_i, n + \ell_i)),$$
$$(1.1.2)$$

and the partial difference inequalities

$$\Delta_m \Delta_n y(m, n) + \sum_{i=1}^{\tau} P_i(m, n, y(g_i(m), h_i(n))) \geq \sum_{i=1}^{\tau} Q_i(m, n, y(g_i(m), h_i(n))),$$
$$(1.1.3)$$

$$\Delta_m \Delta_n y(m, n) + \sum_{i=1}^{\tau} P_i(m, n, y(g_i(m), h_i(n))) \leq \sum_{i=1}^{\tau} Q_i(m, n, y(g_i(m), h_i(n))),$$
$$(1.1.4)$$

where $m \geq m_0$, $n \geq n_0$ in (1.1.1)–(1.1.4), ∇ is the backward difference operator defined by $\nabla y(m) = y(m) - y(m - 1)$, k, ℓ, k_i, ℓ_i, $1 \leq i \leq \tau$ are nonnegative integers, Δ is the forward difference operator defined by $\Delta_m y(m, n) = y(m + 1, n) - y(m, n)$, and $g_i(m)$, $h_i(n)$ ($\in \mathbb{N}$), $1 \leq i \leq \tau$ are deviating arguments satisfying

$$g_i(m) \leq m - \alpha_i \quad \text{and} \quad h_i(n) \leq n - \beta_i \qquad (1.1.5)$$

for some nonnegative integers α_i, β_i, $1 \leq i \leq \tau$. Note that condition (1.1.5) is quite general as it includes not only simple delays (e.g. $g_i(m) = m - \alpha_i$), but also many other types of deviating functions (e.g. $g_i(m) = m - \alpha_i - \left[\frac{m - \alpha_i}{2}\right]$; $g_i(m) = \left[(m - \alpha_i)^\delta\right]$, $0 < \delta \leq 1$).

Throughout, with respect to equation (1.1.1) we shall assume that there exists a function $f : \mathbb{R} \to \mathbb{R}$ and double sequences $\{p(m, n)\}$, $\{p'(m, n)\}$, $\{q(m, n)\}$, $\{q'(m, n)\}$ such that

(A1) for $u \neq 0$, $uf(u) > 0$, $\dfrac{f(u)}{u} \geq \gamma \in (0, \infty)$;

(A2) for $u \neq 0$,

$$p(m, n) \leq \frac{P(m, n, u(m + k, n + \ell))}{f(u(m + k, n + \ell))} \leq p'(m, n),$$

$$q(m, n) \leq \frac{Q(m, n, u(m + k, n + \ell))}{f(u(m + k, n + \ell))} \leq q'(m, n); \quad \text{and}$$

(A3) $\displaystyle\limsup_{m, n \to \infty} [q(m, n) - p'(m, n)] > \dfrac{1}{\gamma} > 0.$

Further, with respect to (1.1.2) for each $1 \leq i \leq \tau$ it is assumed that there exists a function $f_i : \mathbb{R} \to \mathbb{R}$ and double sequences $\{p_i(m, n)\}$, $\{p_i'(m, n)\}$, $\{q_i(m, n)\}$, $\{q_i'(m, n)\}$ such that

(B1) for $u \neq 0$, $uf_i(u) > 0$, $\displaystyle\liminf_{u \to \infty} \dfrac{f_i(u)}{u} = \gamma_i \in (0, \infty)$;

(B2) for $u \neq 0$,

$$p_i(m, n) \leq \frac{P_i(m, n, u(m + k_i, n + \ell_i))}{f_i(u(m + k_i, n + \ell_i))} \leq p_i'(m, n),$$

$$q_i(m, n) \leq \frac{Q_i(m, n, u(m + k_i, n + \ell_i))}{f_i(u(m + k_i, n + \ell_i))} \leq q_i'(m, n); \quad \text{and}$$

(B3) $q_i(m, n) > p_i'(m, n)$ eventually.

With respect to (1.1.3) and (1.1.4), for each $1 \leq i \leq \tau$ we shall assume that there exists a function $f_i : \mathbb{R} \to \mathbb{R}$ and double sequences $\{p_i(m, n)\}$, $\{p_i'(m, n)\}$, $\{q_i(m, n)\}$, $\{q_i'(m, n)\}$ such that

(C1) for $u \neq 0$, $uf_i(u) > 0$, $\dfrac{f_i(u)}{u} \geq \gamma_i \in (0, \infty)$;

(C2) for $u \neq 0$,

$$p_i(m, n) \leq \frac{P_i(m, n, u(g_i(m), h_i(n)))}{f_i(u(g_i(m), h_i(n)))} \leq p_i'(m, n),$$

$$q_i(m, n) \leq \frac{Q_i(m, n, u(g_i(m), h_i(n)))}{f_i(u(g_i(m), h_i(n)))} \leq q_i'(m, n); \quad \text{and}$$

(C3) $q_i(m, n) > p_i'(m, n)$ eventually.

Further, we shall introduce the following notations:

$$\alpha = \min_{1 \le i \le \tau} \alpha_i, \quad \beta = \min_{1 \le i \le \tau} \beta_i, \quad \mu_i(m, n) = q_i(m, n) - p'_i(m, n).$$

By a solution of (1.1.1) (or (1.1.2), (1.1.3), (1.1.4)), we mean a nontrivial double sequence $\{y(m, n)\}$ satisfying (1.1.1) (or (1.1.2), (1.1.3), (1.1.4)) for $m \ge m_0$, $n \ge n_0$. A solution $\{y(m, n)\}$ of (1.1.1) or (1.1.2) is *nondecreasing* (*nonincreasing*) if $\nabla_m y(m, n) \ge (\le) 0$ and $\nabla_n y(m, n) \ge (\le) 0$. A solution $\{y(m, n)\}$ of (1.1.3) or (1.1.4) is *nonincreasing* (*nondecreasing*) if both $\Delta_m y(m, n)$ and $\Delta_n y(m, n)$ are nonpositive (nonnegative). Further, $\{y(m, n)\}$ is *eventually positive* (*negative*) if $y(m, n) > (<) 0$ for all large m and n. A solution of (1.1.1) (or (1.1.2), (1.1.3), (1.1.4)) is said to be *oscillatory* if it is neither eventually positive nor negative, and nonoscillatory otherwise .

In Sections 2 and 3 respectively, we shall offer sufficient conditions for the nonexistence of eventually positive (negative) as well as nondecreasing (nonincreasing) solutions of equations (1.1.1) and (1.1.2). The nonexistence criteria of eventually positive (negative) and nonincreasing (nondecreasing) solutions of (1.1.3) ((1.1.4)) are presented in Section 4. Finally, to illustrate the results obtained, we also include some examples in Section 5. We remark that our nonexistence criteria certainly extend/complement the oscillation results of several authors [22–28, 36–38].

1.2 Nonexistence criteria for (1.1.1)

Throughout this section we shall use the following notations:

$$\alpha = \min\{k, \ell\}, \quad \beta = \max\{k, \ell\}, \quad \mu(m, n) = q(m, n) - p'(m, n).$$

Further, we define

$$E = \{r > 0 \mid 1 - r[\gamma\mu(m, n) - 1] > 0 \text{ eventually}\}.$$

Theorem 1.2.1. Suppose that there exist integers $M \ge m_0$, $N \ge n_0$ such that

$$\sup_{r \in E, m \ge M, n \ge N} r \prod_{i=1}^{k} \prod_{j=1}^{\ell} \{1 - r[\gamma\mu(m+i, n+j) - 1]\}^{1/\alpha} < 1. \qquad (1.2.1)$$

Then, equation (1.1.1) has no eventually positive (negative) and nondecreasing (nonincreasing) solution.

Proof. (a) Let $\{y(m, n)\}$ be an eventually positive and nondecreasing solution of (1.1.1). We define

$$S = \{r > 0 \mid -\nabla_m \nabla_n y(m,n) + y(m-1,n-1) + r[\gamma\mu(m,n)-1]y(m,n)$$

$$\leq 0 \text{ eventually}\}. \tag{1.2.2}$$

First, we shall show that the set S is nonempty. For this, in view of (A1) - (A3), we find for large m and n,

$$-\nabla_m \nabla_n y(m,n) + y(m-1,n-1) + [\gamma\mu(m,n)-1]y(m,n)$$

$$\leq -\mu(m,n)f(y(m+k,n+\ell)) + y(m-1,n-1) + [\gamma\mu(m,n)-1]y(m,n)$$

$$\leq -\mu(m,n)\gamma y(m+k,n+\ell) + y(m-1,n-1) + [\gamma\mu(m,n)-1]y(m,n)$$

$$\leq \{-\mu(m,n)\gamma + 1 + [\gamma\mu(m,n)-1]\}y(m+k,n+\ell) = 0 \tag{1.2.3}$$

where we have used the monotone property of $y(m,n)$ in the last inequality. It follows from (1.2.3) that $1 \in S$ and so S is nonempty.

Next, we shall prove that S is bounded. For this, let $r \in S$. Then, from definition (1.2.2) it is clear that

$$y(m,n-1) + y(m-1,n) - \{1 - r[\gamma\mu(m,n)-1]\}y(m,n) \leq 0 \tag{1.2.4}$$

eventually. Hence, we must have $1 - r[\gamma\mu(m,n)-1] > 0$ eventually. So $r \in E$ and we see that $S \subseteq E$. For any $r \in E$, in view of (A3) we have

$$r < \frac{1}{\gamma\mu(m,n)-1} < \infty.$$

Thus, E is bounded, and so is S.

Now, let $\bar{r} \in S$. Then, from (1.2.4) we have for large m and n,

$$y(m,n-1) \leq \{1 - \bar{r}[\gamma\mu(m,n)-1]\}y(m,n) \tag{1.2.5}$$

and

$$y(m-1,n) \leq \{1 - \bar{r}[\gamma\mu(m,n)-1]\}y(m,n). \tag{1.2.6}$$

Repeated application of (1.2.5) and (1.2.6) respectively leads to

$$y(m,n) \leq \prod_{j=1}^{\ell}\{1 - \bar{r}[\gamma\mu(m,n+j)-1]\}y(m,n+\ell) \tag{1.2.7}$$

and

$$y(m,n) \leq \prod_{i=1}^{k}\{1 - \bar{r}[\gamma\mu(m+i,n)-1]\}y(m+k,n). \tag{1.2.8}$$

Noting that $y(m,n)$ is nondecreasing and also (1.2.7), we get

$$
\begin{aligned}
[y(m,n)]^k &\le \prod_{i=1}^{k} y(m+i,n) \\
&\le \prod_{i=1}^{k}\prod_{j=1}^{\ell}\{1 - \bar{r}[\gamma\mu(m+i,n+j)-1]\}y(m+i,n+\ell) \\
&\le \prod_{i=1}^{k}\prod_{j=1}^{\ell}\{1 - \bar{r}[\gamma\mu(m+i,n+j)-1]\}[y(m+k,n+\ell)]^k \qquad (1.2.9)
\end{aligned}
$$

or

$$
y(m,n) \le \prod_{i=1}^{k}\prod_{j=1}^{\ell}\{1 - \bar{r}[\gamma\mu(m+i,n+j)-1]\}^{1/k}y(m+k,n+\ell). \qquad (1.2.10)
$$

Similarly, on using (1.2.8) we find

$$
\begin{aligned}
[y(m,n)]^{\ell} &\le \prod_{j=1}^{\ell} y(m,n+j) \\
&\le \prod_{j=1}^{\ell}\prod_{i=1}^{k}\{1 - \bar{r}[\gamma\mu(m+i,n+j)-1]\}y(m+k,n+j) \\
&\le \prod_{j=1}^{\ell}\prod_{i=1}^{k}\{1 - \bar{r}[\gamma\mu(m+i,n+j)-1]\}[y(m+k,n+\ell)]^{\ell}
\end{aligned}
$$

which is the same as

$$
y(m,n) \le \prod_{i=1}^{k}\prod_{j=1}^{\ell}\{1 - \bar{r}[\gamma\mu(m+i,n+j)-1]\}^{1/\ell}y(m+k,n+\ell). \qquad (1.2.11)
$$

Taking note of the fact that $1 - \bar{r}[\gamma\mu(m+i,n+j)-1]$ is positive and less than 1, we combine (1.2.10) and (1.2.11) to obtain

$$
\begin{aligned}
y(m,n) \le \min\Bigg\{ &\prod_{i=1}^{k}\prod_{j=1}^{\ell}\{1 - \bar{r}[\gamma\mu(m+i,n+j)-1]\}^{1/k}, \\
&\prod_{i=1}^{k}\prod_{j=1}^{\ell}\{1 - \bar{r}[\gamma\mu(m+i,n+j)-1]\}^{1/\ell}\Bigg\} y(m+k,n+\ell) \\
= &\prod_{i=1}^{k}\prod_{j=1}^{\ell}\{1 - \bar{r}[\gamma\mu(m+i,n+j)-1]\}^{1/\alpha}y(m+k,n+\ell). \qquad (1.2.12)
\end{aligned}
$$

Now, in view of (A1) - (A3), we find

$$-\nabla_m \nabla_n y(m,n) + y(m-1, n-1)$$
$$\leq -\mu(m,n)f(y(m+k, n+\ell)) + y(m-1, n-1)$$
$$\leq -\mu(m,n)\ \gamma y(m+k, n+\ell) + y(m-1, n-1)$$
$$\leq [1 - \gamma\mu(m,n)]y(m+k, n+\ell)$$

or

$$-\nabla_m \nabla_n y(m,n) + y(m-1, n-1) + [\gamma\mu(m,n) - 1]y(m+k, n+\ell) \leq 0. \quad (1.2.13)$$

Applying (1.2.12) in (1.2.13) gives

$$-\nabla_m \nabla_n y(m,n) + y(m-1, n-1) + [\gamma\mu(m,n) - 1]\times$$
$$\prod_{i=1}^{k}\prod_{j=1}^{\ell}\{1 - \bar{r}[\gamma\mu(m+i, n+j) - 1]\}^{-1/\alpha}y(m,n) \leq 0 \quad (1.2.14)$$

which implies

$$-\nabla_m \nabla_n y(m,n) + y(m-1, n-1) + [\gamma\mu(m,n) - 1]\times$$
$$\left\{\sup_{m \geq M, n \geq N} \prod_{i=1}^{k}\prod_{j=1}^{\ell}[1 - \bar{r}(\gamma\mu(m+i, n+j) - 1)]\right\}^{-1/\alpha} y(m,n) \leq 0. \quad (1.2.15)$$

It follows immediately from the above inequality that

$$\left\{\sup_{m \geq M, n \geq N} \prod_{i=1}^{k}\prod_{j=1}^{\ell}[1 - \bar{r}(\gamma\mu(m+i, n+j) - 1)]\right\}^{-1/\alpha} \in S. \quad (1.2.16)$$

On the other hand, condition (1.2.1) implies the existence of a $c_1 \in (0,1)$ such that

$$\sup_{r \in E, m \geq M, n \geq N} r\prod_{i=1}^{k}\prod_{j=1}^{\ell}\{1 - r[\gamma\mu(m+i, n+j) - 1]\}^{1/\alpha} \leq c_1 < 1.$$

In particular, this leads to (when $r = \bar{r}$)

$$\left\{\sup_{m \geq M, n \geq N} \prod_{i=1}^{k}\prod_{j=1}^{\ell}[1 - \bar{r}(\gamma\mu(m+i, n+j) - 1)]\right\}^{-1/\alpha} \geq \frac{\bar{r}}{c_1}. \quad (1.2.17)$$

Since $\bar{r} \in S$ and $r' \leq \bar{r}$ imply that $r' \in S$, it follows from (1.2.16) and (1.2.17) that $\bar{r}/c_1 \in S$.

Repeating the above arguments with $\bar{r}$ replaced by $\bar{r}/c_1$, we get $\bar{r}/(c_1 c_2) \in S$ where $c_2 \in (0, 1)$. Continuing in this way, we obtain

$$\bar{r} \left(\prod_{i=1}^{\infty} c_i \right)^{-1} \in S \tag{1.2.18}$$

where $c_i \in (0, 1)$. This contradicts the boundedness of S.

(b) Let $\{y(m, n)\}$ be an eventually negative and nonincreasing solution of (1.1.1). Here, we define

$$S' = \{r > 0 \mid -\nabla_m \nabla_n y(m, n) + y(m - 1, n - 1) + r[\gamma\mu(m, n) - 1]y(m, n)$$

$$\geq 0 \text{ eventually}\}.$$

We shall use the equation number $(1.2.\bullet)'$ to denote $(1.2.\bullet)$ with the inequality sign reversed. Using (A1) - (A3), we obtain $(1.2.3)'$. Hence, $1 \in S'$. Next, let $r \in S'$. From the definition of S' we get $(1.2.4)'$. Since $\{y(m, n)\}$ is eventually negative, this implies that $1 - r[\gamma\mu(m, n) - 1] > 0$ eventually. Thus, $r \in E$ and so $S' \subseteq E$. It follows that S' is bounded (as E is).

Now, let $\bar{r} \in S'$. Then, we obtain $(1.2.5)'$ - $(1.2.8)'$. First, suppose that k is odd. On using $(1.2.7)'$, we obtain $(1.2.9)'$ which is equivalent to $(1.2.10)'$. Next, suppose that k is even. Then, an application of $(1.2.7)'$ gives exactly $(1.2.9)$ from which we find

$$|y(m, n)| \leq \prod_{i=1}^{k} \prod_{j=1}^{\ell} \{1 - \bar{r}[\gamma\mu(m + i, n + j) - 1]\}^{1/k} |y(m + k, n + \ell)|.$$

Since $|y| = -y$, the above inequality is exactly the same as $(1.2.10)'$. Hence, in both cases we get $(1.2.10)'$. Likewise, we also have $(1.2.11)'$. Combining $(1.2.10)'$ and $(1.2.11)'$, we get $(1.2.12)'$.

Using (A1) - (A3), we obtain $(1.2.13)'$. An application of $(1.2.12)'$ then leads to $(1.2.14)'$ and $(1.2.15)'$. Thus, we conclude that $(1.2.16)$ holds with S replaced by S'. The rest of the proof follows as before. $\square$

Corollary 1.2.2. Suppose that $k, \ell \geq 1$ and

$$\liminf_{m,n \to \infty} \frac{1}{k\ell} \sum_{i=1}^{k} \sum_{j=1}^{\ell} [\gamma\mu(m + i, n + j) - 1] > \frac{\beta^\beta}{(1 + \beta)^{1+\beta}}. \tag{1.2.19}$$

Then, the conclusion of Theorem 1.2.1 holds.

Proof. Using the well-known arithmetic-geometric mean inequality, we have

$$\prod_{i=1}^{k}\prod_{j=1}^{\ell}\{1-r[\gamma\mu(m+i,n+j)-1]\}^{1/k\ell}$$

$$\leq \frac{1}{k\ell}\sum_{i=1}^{k}\sum_{j=1}^{\ell}\{1-r[\gamma\mu(m+i,n+j)-1]\}$$

$$= 1-\frac{r}{k\ell}\sum_{i=1}^{k}\sum_{j=1}^{\ell}[\gamma\mu(m+i,n+j)-1].$$

Noting that $\beta/(k\ell)=1/\alpha$, it follows from the above inequality that

$$\prod_{i=1}^{k}\prod_{j=1}^{\ell}\{1-r[\gamma\mu(m+i,n+j)-1]\}^{1/\alpha} \leq \left\{1-\frac{r}{k\ell}\sum_{i=1}^{k}\sum_{j=1}^{\ell}[\gamma\mu(m+i,n+j)-1]\right\}^{\beta}. \tag{1.2.20}$$

Now, let

$$\delta = \frac{1}{k\ell}\sum_{i=1}^{k}\sum_{j=1}^{\ell}[\gamma\mu(m+i,n+j)-1] \tag{1.2.21}$$

and

$$g(u) = u(1-u\delta)^{\beta}.$$

It is computed that

$$\max_{u>0} g(u) = g\left(\frac{1}{\delta(1+\beta)}\right) = \frac{\beta^{\beta}}{(1+\beta)^{1+\beta}}\,\delta^{-1} < 1 \tag{1.2.22}$$

where we have used (1.2.19) in the last inequality. Coupling (1.2.22) and (1.2.20), we get (1.2.1) and so the conclusion follows. □

Theorem 1.2.3. Suppose that there exist integers $M \geq m_0$, $N \geq n_0$ such that if $\ell \geq k$,

$$\sup_{r\in E, m\geq M, n\geq N} r\prod_{i=1}^{k}\{1-r[\gamma\mu(m+i,n+i)-1]\}\times$$

$$\prod_{j=k+1}^{\ell}\{1-r[\gamma\mu(m+k,n+j)-1]\} < 2^k; \tag{1.2.23}$$

and if $\ell < k$,

$$\sup_{r\in E, m\geq M, n\geq N} r\prod_{i=1}^{\ell}\{1-r[\gamma\mu(m+i,n+i)-1]\}\times$$

$$\prod_{j=\ell+1}^{k} \{1 - r[\gamma\mu(m+j, n+\ell) - 1]\} < 2^{\ell}. \tag{1.2.24}$$

Then, the conclusion of Theorem 1.2.1 holds.

Proof. Let $\{y(m,n)\}$ be an eventually positive and nondecreasing solution of (1.1.1). We shall consider only this case as the proof is similar when $\{y(m,n)\}$ is an eventually negative and nonincreasing solution of (1.1.1). Let $\bar{r} \in S$ where S is defined in (1.2.2). As before we have (1.2.4) which we use to get

$$2\, y(m,n) \leq y(m+1, n) + y(m, n+1) \leq \{1 - \bar{r}[\gamma\mu(m+1, n+1) - 1]\}y(m+1, n+1)$$

or

$$y(m,n) \leq \frac{1}{2}\{1 - \bar{r}[\gamma\mu(m+1, n+1) - 1]\}y(m+1, n+1). \tag{1.2.25}$$

Repeated application of (1.2.25) yields

$$y(m,n) \leq \left(\frac{1}{2}\right)^{\alpha} \prod_{i=1}^{\alpha} \{1 - \bar{r}[\gamma\mu(m+i, n+i) - 1]\}y(m+\alpha, n+\alpha). \tag{1.2.26}$$

Suppose that $\ell \geq k$, i.e., $\alpha = k$. The proof is similar if $\ell < k$. From (1.2.5), we find

$$y(m+k, n+k) \leq \prod_{j=k+1}^{\ell} \{1 - \bar{r}(\gamma\mu(m+k, n+j) - 1)\}y(m+k, n+\ell)$$

which we use in (1.2.26) to obtain

$$y(m,n) \leq \left(\frac{1}{2}\right)^{k} \prod_{i=1}^{k} \{1 - \bar{r}[\gamma\mu(m+i, n+i) - 1]\} \times$$

$$\prod_{j=k+1}^{\ell} \{1 - \bar{r}[\gamma\mu(m+k, n+j) - 1]\}y(m+k, n+\ell). \tag{1.2.27}$$

In view of (A1) - (A3), we have (1.2.13) as earlier. Applying (1.2.27) in (1.2.13), we get

$$-\nabla_m \nabla_n y(m,n) + y(m-1, n-1) + [\gamma\mu(m,n) - 1]y(m,n) \times$$

$$2^k \left\{ \prod_{i=1}^{k} [1 - \bar{r}(\gamma\mu(m+i, n+i) - 1)] \prod_{j=k+1}^{\ell} [1 - \bar{r}(\gamma\mu(m+k, n+j) - 1)] \right\}^{-1} \leq 0$$

which implies

$$2^k \left\{ \sup_{m \geq M, n \geq N} \prod_{i=1}^{k} [1 - \bar{r}(\gamma \mu(m+i, n+i) - 1)] \times \right.$$

$$\left. \prod_{j=k+1}^{\ell} [1 - \bar{r}(\gamma \mu(m+k, n+j) - 1)] \right\}^{-1} \in S. \qquad (1.2.28)$$

On the other hand, condition (1.2.23) ensures the existence of a $c_1 \in (0,1)$ such that

$$\bar{r} \sup_{m \geq M, n \geq N} \prod_{i=1}^{k} \{1 - \bar{r}[\gamma \mu(m+i, n+i) - 1]\} \prod_{j=k+1}^{\ell} \{1 - \bar{r}[\gamma \mu(m+k, n+j) - 1]\} \leq c_1 \, 2^k$$

or

$$2^k \left\{ \sup_{m \geq M, n \geq N} \prod_{i=1}^{k} [1 - \bar{r}(\gamma \mu(m+i, n+i) - 1)] \times \right.$$

$$\left. \prod_{j=k+1}^{\ell} [1 - \bar{r}(\gamma \mu(m+k, n+j) - 1)] \right\}^{-1} \geq \frac{\bar{r}}{c_1}. \qquad (1.2.29)$$

It now follows from (1.2.28) and (1.2.29) that $\bar{r}/c_1 \in S$.

Repeating the above procedure with $\bar{r}$ replaced by $\bar{r}/c_1$, we get $\bar{r}/(c_1 c_2) \in S$ where $c_2 \in (0,1)$. Continuing in this manner, we obtain (1.2.18) which contradicts the fact that S is bounded. □

Corollary 1.2.4. Suppose that

$$\liminf_{m,n \to \infty} \mu(m,n) > \frac{1}{\gamma} \left[\frac{\beta^\beta}{2^\alpha (1+\beta)^{1+\beta}} + 1 \right]. \qquad (1.2.30)$$

Then, the conclusion of Theorem 1.2.1 holds.

Proof. Let

$$\delta = \liminf_{m,n \to \infty} [\gamma \mu(m,n) - 1]$$

and

$$g(u) = u(1 - u\delta)^\beta.$$

In view of (1.2.30), we find that

$$\max_{u>0} g(u) = \frac{\beta^\beta}{(1+\beta)^{1+\beta}} \, \delta^{-1} < 2^\alpha.$$

Hence, it is immediate that

$$\sup_{r \in E} r(1 - r\delta)^{\beta} < 2^{\alpha}. \tag{1.2.31}$$

Since (1.2.31) implies (1.2.23) and (1.2.24), the conclusion now follows from Theorem 1.2.3. $\qquad\square$

Theorem 1.2.5. Suppose that there exist integers $M \geq m_0$, $N \geq n_0$ such that

$$\sup_{r \in E, m \geq M, n \geq N} r \prod_{i=1}^{k} \{1 - r[\gamma\mu(m + i, n) - 1]\} \prod_{j=1}^{\ell} \{1 - r[\gamma\mu(m + k, n + j) - 1]\} < 1. \tag{1.2.32}$$

Then, the conclusion of Theorem 1.2.1 holds.

Proof. Once again we shall only consider the case when $\{y(m, n)\}$ is an eventually positive and nondecreasing solution of (1.1.1). Let $\bar{r} \in S$ where S is defined in (1.2.2). Then, from (1.2.7) we get

$$y(m + k, n) \leq \prod_{j=1}^{\ell} \{1 - \bar{r}[\gamma\mu(m + k, n + j) - 1]\} y(m + k, n + \ell). \tag{1.2.33}$$

A combination of (1.2.8) and (1.2.33) yields

$$y(m, n) \leq \prod_{i=1}^{k} \{1 - \bar{r}[\gamma\mu(m + i, n) - 1]\} \prod_{j=1}^{\ell} \{1 - \bar{r}[\gamma\mu(m + k, n + j) - 1]\} y(m + k, n + \ell)$$

which we use in (1.2.13) to provide

$$-\nabla_m \nabla_n y(m, n) + y(m - 1, n - 1) + [\gamma\mu(m, n) - 1]y(m, n) \times$$

$$\left\{ \prod_{i=1}^{k} [1 - \bar{r}(\gamma\mu(m + i, n) - 1)] \prod_{j=1}^{\ell} [1 - \bar{r}(\gamma\mu(m + k, n + j) - 1)] \right\}^{-1} \leq 0. \tag{1.2.34}$$

It is now immediate from (1.2.34) that

$$\left\{ \sup_{m \geq M, n \geq N} \prod_{i=1}^{k} [1 - \bar{r}(\gamma\mu(m + i, n) - 1)] \prod_{j=1}^{\ell} [1 - \bar{r}(\gamma\mu(m + k, n + j) - 1)] \right\}^{-1} \in S. \tag{1.2.35}$$

On the other hand, in view of (1.2.32) there exists a $c_1 \in (0, 1)$ such that

$$\sup_{m \geq M, n \geq N} \bar{r} \prod_{i=1}^{k} \{1 - \bar{r}[\gamma\mu(m + i, n) - 1]\} \prod_{j=1}^{\ell} \{1 - \bar{r}[\gamma\mu(m + k, n + j) - 1]\} \leq c_1.$$

Together with (1.2.35), this implies that $\bar{r}/c_1 \in S$. Then, using a similar argument as in Theorem 1.2.1, we obtain (1.2.18) which is a contradiction to the boundedness of S. □

Corollary 1.2.6. Suppose that

$$\mu(m,n) \geq c > \frac{1}{\gamma}\left[\frac{(k+\ell)^{k+\ell}}{(k+\ell+1)^{k+\ell+1}} + 1\right]. \tag{1.2.36}$$

Then, the conclusion of Theorem 1.2.1 holds.

Proof. Let

$$g(u) = u[1 - u(\gamma c - 1)]^{k+\ell}.$$

Then,

$$\max_{u>0} g(u) = \frac{(k+\ell)^{k+\ell}}{(\gamma c - 1)(k+\ell+1)^{k+\ell+1}} < 1 \tag{1.2.37}$$

where we have used (1.2.36) in the last inequality. From (1.2.37) it follows that

$$\sup_{r \in E} r[1 - r(\gamma c - 1)]^{k+\ell} < 1. \tag{1.2.38}$$

In view of (1.2.38), we find

$$\sup_{r \in E, m \geq M, n \geq N} r \prod_{i=1}^{k}\{1 - r[\gamma\mu(m+i,n) - 1]\} \prod_{j=1}^{\ell}\{1 - r[\gamma\mu(m+k,n+j) - 1]\}$$

$$\leq \sup_{r \in E} r[1 - r(\gamma c - 1)]^{k+\ell} < 1.$$

The conclusion now follows immediately from Theorem 1.2.5. □

1.3   Nonexistence criteria for (1.1.2)

Lemma 1.3.1. [32] The following identity is true

$$\sum_{i=m-k_0}^{m} \sum_{j=n-\ell_0}^{n} [y(i-1,j) + y(i,j-1) - y(i,j)]$$

$$= \sum_{i=m-k_0}^{m-1} \sum_{j=n-\ell_0}^{n} y(i,j-1) + \sum_{j=n-\ell_0}^{n} y(m-k_0-1,j) + y(m,n-\ell_0-1) - y(m,n).$$

Theorem 1.3.2. Suppose that for each $1 \le i \le \tau$, f_i is nondecreasing;

$$\liminf_{m,n\to\infty} p_i'(m,n) = p_i', \qquad \liminf_{m,n\to\infty} q_i(m,n) = q_i, \qquad q_i > p_i'; \tag{1.3.1}$$

and

$$\sum_{i=1}^{\tau}(q_i - p_i')\gamma_i \, \frac{(\alpha_i+1)^{\alpha_i+1}}{\alpha_i^{\alpha_i}} > 1 \tag{1.3.2}$$

where $\alpha_i = \min\{k_i, \ell_i\}$, $1 \le i \le \tau$. Then, equation (1.1.2) has no eventually positive (negative) and nondecreasing (nonincreasing) solution.

Proof. Let $\{y(m,n)\}$ be an eventually positive and nondecreasing solution of (1.1.2). We shall only consider this case as the proof is similar if $\{y(m,n)\}$ is eventually negative and nonincreasing. First, we claim that $\lim_{m,n\to\infty} y(m,n) = \infty$. Suppose the contrary, i.e., $\lim_{m,n\to\infty} y(m,n) = c < \infty$. In view of (B1) - (B3), from (1.1.2) we have

$$\nabla_m\nabla_n y(m,n) \ge \sum_{i=1}^{\tau}[q_i(m,n) - p_i'(m,n)]f_i(y(m+k_i,n+\ell_i)). \tag{1.3.3}$$

Taking limit infimum in (1.3.3) and noting that $\nabla_m\nabla_n y(m,n) = y(m,n) - y(m,n-1) - y(m-1,n) + y(m-1,n-1)$, we get

$$0 \ge \sum_{i=1}^{\tau}(q_i - p_i')f_i(c) > 0$$

which is a contradiction. Hence, $\lim_{m,n\to\infty} y(m,n) = \infty$.

Next, writing $b(m,n) = y(m,n)/y(m-1,n-1)$ (≥ 1), we have

$$
\begin{aligned}
-\frac{\nabla_m\nabla_n y(m,n)}{y(m,n)} &= \frac{y(m,n-1) + y(m-1,n) - y(m-1,n-1)}{y(m,n)} - 1 \\
&\ge \frac{2y(m-1,n-1) - y(m-1,n-1)}{y(m,n)} - 1 \\
&= \frac{1}{b(m,n)} - 1.
\end{aligned} \tag{1.3.4}
$$

Since f_i is nondecreasing, inequality (1.3.4) provides

$$
\begin{aligned}
\frac{1}{b(m,n)} - 1 &\le -\frac{\nabla_m\nabla_n y(m,n)}{y(m,n)} \\
&\le \sum_{i=1}^{\tau}[p_i'(m,n) - q_i(m,n)]\frac{f_i(y(m+k_i,n+\ell_i))}{y(m,n)} \\
&\le \sum_{i=1}^{\tau}[p_i'(m,n) - q_i(m,n)]\frac{f_i(y(m+\alpha_i,n+\alpha_i))}{y(m,n)}
\end{aligned}
$$

$$= \sum_{i=1}^{\tau} [p_i'(m,n) - q_i(m,n)] \prod_{s=1}^{\alpha_i} b(m+s, n+s) \, \frac{f_i(y(m+\alpha_i, n+\alpha_i))}{y(m+\alpha_i, n+\alpha_i)}. \quad (1.3.5)$$

It follows from (1.3.5) that $b(m,n)$ is bounded, for otherwise we get the contradiction $-1 \le -\infty$.

Let $d = \liminf_{m,n\to\infty} b(m,n)$. Then, $d \in [1,\infty)$. Taking limit infimum in (1.3.5) gives

$$\frac{1}{d} - 1 \le \sum_{i=1}^{\tau} (p_i' - q_i) d^{\alpha_i} \gamma_i \, (< 0) \quad (1.3.6)$$

where we have used (1.3.1). From (1.3.6) we note that $d > 1$. Rewriting (1.3.6), we obtain

$$\sum_{i=1}^{\tau} (q_i - p_i') \gamma_i \frac{d^{\alpha_i+1}}{d-1} \le 1. \quad (1.3.7)$$

Let $F(d) = d^{\alpha_i+1}/(d-1)$. Then,

$$\min_{d>1} F(d) = F\left(\frac{\alpha_i+1}{\alpha_i}\right) = \frac{(\alpha_i+1)^{\alpha_i+1}}{\alpha_i^{\alpha_i}}. \quad (1.3.8)$$

Thus, applying (1.3.8) in (1.3.7) leads to

$$\sum_{i=1}^{\tau} (q_i - p_i') \gamma_i \frac{(\alpha_i+1)^{\alpha_i+1}}{\alpha_i^{\alpha_i}} \le 1$$

which contradicts (1.3.2). □

Theorem 1.3.3. Suppose that for each $1 \le i \le \tau$, f_i is nondecreasing and

$$\limsup_{m,n\to\infty} \sum_{s=1}^{\tau} \gamma_s \sum_{i=m-k'}^{m} \sum_{j=n-\ell'}^{n} [q_s(i,j) - p_s'(i,j)] > \ell' + 2 \quad (1.3.9)$$

where $k' = \min_{1\le i\le\tau} k_i$, $\ell' = \min_{1\le i\le\tau} \ell_i$. Then, the conclusion of Theorem 1.3.2 holds.

Proof. Once again we let $\{y(m,n)\}$ be an eventually positive and nondecreasing solution of (1.1.2). Summing (1.1.2) and applying Lemma 1.3.1, we get

$$\sum_{i=m-k'}^{m} \sum_{j=n-\ell'}^{n} \sum_{s=1}^{\tau} [Q_s(i,j,y(i+k_s,j+\ell_s)) - P_s(i,j,y(i+k_s,j+\ell_s))]$$

$$= \sum_{i=m-k'}^{m} \sum_{j=n-\ell'}^{n} \nabla_i \nabla_j y(i,j)$$

$$= \sum_{i=m-k'}^{m} \sum_{j=n-\ell'}^{n} [y(i,j) - y(i-1,j) - y(i,j-1) + y(i-1,j-1)]$$

$$= y(m,n) - y(m, n-\ell'-1) - \sum_{j=n-\ell'}^{n} y(m-k'-1, j) - \sum_{i=m-k'}^{m-1} \sum_{j=n-\ell'}^{n} y(i, j-1)$$

$$+ \sum_{i=m-k'}^{m-1} \sum_{j=n-\ell'}^{n} y(i-1, j-1) + \sum_{j=n-\ell'}^{n} y(m-1, j-1)$$

$$\le y(m,n) + \sum_{j=n-\ell'}^{n} y(m-1, j-1)$$

$$\le y(m,n) + (\ell'+1)y(m,n) = (\ell'+2)y(m,n). \tag{1.3.10}$$

Using the monotone property of f_i, $1 \le i \le \tau$, it follows from (1.3.10) that

$$(\ell'+2)y(m,n) \ge \sum_{i=m-k'}^{m} \sum_{j=n-\ell'}^{n} \sum_{s=1}^{\tau} [q_s(i,j) - p'_s(i,j)] f_s(y(i+k_s, j+\ell_s))$$

$$\ge \sum_{i=m-k'}^{m} \sum_{j=n-\ell'}^{n} \sum_{s=1}^{\tau} [q_s(i,j) - p'_s(i,j)] f_s(y(m,n))$$

or equivalently

$$\sum_{s=1}^{\tau} \frac{f_s(y(m,n))}{y(m,n)} \sum_{i=m-k'}^{m} \sum_{j=n-\ell'}^{n} [q_s(i,j) - p'_s(i,j)] \le \ell' + 2. \tag{1.3.11}$$

Taking limit supremum in (1.3.11), we immediately get a contradiction to (1.3.9). $\square$

1.4   Nonexistence criteria for (1.1.3) and (1.1.4)

We shall state some lemmas [32] before presenting the main results.

Lemma 1.4.1. Let $\{y(m,n)\}$ be any eventually positive (negative) sequence. Suppose that there exist $\lambda \in \mathbb{R}$ such that for all large m and n,

$$y(m+1,n) + y(m, n+1) - \lambda y(m,n) \le (\ge) \, 0. \tag{1.4.1}$$

Then, for $\alpha, \beta \ge 0$ and large m and n,

$$y(m-\alpha, n-\beta) \ge (\le) \, 2^r \lambda^{-\alpha-\beta} y(m,n) \tag{1.4.2}$$

where $r = \min\{\alpha, \beta\}$.

Lemma 1.4.2. The following identity holds for $M \leq m$, $N \leq n$

$$\sum_{i=M}^{m} \sum_{j=N}^{n} -\Delta_i \Delta_j y(i,j) = y(m+1, N) - y(M, N) + y(M, n+1) - y(m+1, n+1).$$

$$(1.4.3)$$

Remark 1.4.3. It is clear from (1.4.3) that if the sequence $\{y(i,j)\}$ is nonincreasing (nondecreasing), then

$$\sum_{i=M}^{m} \sum_{j=N}^{n} -\Delta_i \Delta_j y(i,j) \geq (\leq) \ y(m+1, N) - y(M, N) \qquad (1.4.4)$$

and also

$$\sum_{i=M}^{m} \sum_{j=N}^{n} -\Delta_i \Delta_j y(i,j) \geq (\leq) \ y(M, n+1) - y(M, N). \qquad (1.4.5)$$

Lemma 1.4.4. Let $\{y(m,n)\}$ be an eventually positive (negative) and nonincreasing (nondecreasing) solution of (1.1.3) ((1.1.4)). Suppose that

$$\sum_{m=M}^{\infty} \sum_{n=N}^{\infty} \sum_{i=1}^{\tau} \gamma_i \mu_i(m,n) = \infty. \qquad (1.4.6)$$

Then, $y(m,n)$ tends to zero as m and n tend to infinity.

Lemma 1.4.5. Let $\{y(m,n)\}$ be an eventually positive (negative) and nonincreasing (nondecreasing) solution of (1.1.3) ((1.1.4)). Suppose that there exists $b > 0$ such that for sufficiently large M and N,

$$\sum_{m=M}^{K} \sum_{n=N}^{L} \sum_{i=1}^{\tau} \gamma_i \mu_i(m,n) \geq b. \qquad (1.4.7)$$

Then,

$$y(M, N) \geq (\leq) \ by(K - \alpha, L - \beta) + y(K + 1, N) \qquad (1.4.8)$$

and

$$y(M, N) \geq (\leq) \ by(K - \alpha, L - \beta) + y(M, L + 1). \qquad (1.4.9)$$

Lemma 1.4.6. Let $\alpha, \beta > 0$ and $\{y(m,n)\}$ be an eventually positive (negative) and nonincreasing (nondecreasing) solution of (1.1.3) ((1.1.4)). Suppose that there exists $b \in (0, 1/2]$ such that for all large M and N,

$$\sum_{m=M-\alpha}^{M-1} \sum_{n=N-\beta}^{N-1} \sum_{i=1}^{\tau} \gamma_i \mu_i(m,n) \geq b. \qquad (1.4.10)$$

Then, for all large s and t,

$$y(s - \alpha, t - \beta) \leq (\geq) \; c^2 y(s + 1, t + 1) \qquad (1.4.11)$$

where $c = 2b^{-2}(1 - b + \sqrt{1 - 2b})$.

Lemma 1.4.7. Let $\alpha, \beta > 0$ and $\{y(m, n)\}$ be an eventually positive (negative) and nonincreasing (nondecreasing) solution of (1.1.3) ((1.1.4)). Suppose that there exists $b > 0$ such that (1.4.10) holds for all large M and N. Then, for all large s and t,

$$y(s - \alpha, t - \beta) \leq (\geq) \; \frac{16}{b^4} \, y(s + 1, t + 1). \qquad (1.4.12)$$

Lemma 1.4.8. Let $\beta = 0$, $\alpha > 0$ and $\{y(m, n)\}$ be an eventually positive (negative) and nonincreasing (nondecreasing) solution of (1.1.3) ((1.1.4)). Suppose that there exists $b \in (0, 1/2]$ such that for all large M and N,

$$\sum_{m=M-\alpha}^{M-1} \sum_{i=1}^{\tau} \gamma_i \mu_i(m, N) \geq b. \qquad (1.4.13)$$

Then, for all large s and t,

$$y(s - \alpha, t) \leq (\geq) \; cy(s + 1, t) \qquad (1.4.14)$$

where c is defined in Lemma 1.4.6.

Lemma 1.4.9. Let $\alpha = 0$, $\beta > 0$ and $\{y(m, n)\}$ be an eventually positive (negative) and nonincreasing (nondecreasing) solution of (1.1.3) ((1.1.4)). Suppose that there exists $b \in (0, 1/2]$ such that for all large M and N,

$$\sum_{n=N-\beta}^{N-1} \sum_{i=1}^{\tau} \gamma_i \mu_i(M, n) \geq b. \qquad (1.4.15)$$

Then, for all large s and t,

$$y(s, t - \beta) \leq (\geq) \; cy(s, t + 1) \qquad (1.4.16)$$

where c is defined in Lemma 1.4.6.

Lemma 1.4.10. Let $\beta = 0$, $\alpha > 0$ and $\{y(m, n)\}$ be an eventually positive (negative) and nonincreasing (nondecreasing) solution of (1.1.3) ((1.1.4)). Suppose that there exists $b > 0$ such that (1.4.13) holds for all large M and N. Then, for all large s and t,

$$y(s - \alpha, t) \leq (\geq) \; \frac{4}{b^2} \, y(s + 1, t). \qquad (1.4.17)$$

Lemma 1.4.11. Let $\alpha = 0$, $\beta > 0$ and $\{y(m,n)\}$ be an eventually positive (negative) and nonincreasing (nondecreasing) solution of (1.1.3) ((1.1.4)). Suppose that there exists $b > 0$ such that (1.4.15) holds for all large M and N. Then, for all large s and t,

$$y(s, t - \beta) \le (\ge) \frac{4}{b^2} \, y(s, t + 1). \tag{1.4.18}$$

Theorem 1.4.12. Let $\alpha + \beta > 0$. Suppose that for each $1 \le i \le \tau$, $\mu_i(m,n) \ge \rho_i > 0$ eventually and

$$d = \frac{\alpha + \beta + 1}{\alpha + \beta} \left[2^r(\alpha + \beta) \sum_{i=1}^{\tau} \rho_i \gamma_i \right]^{1/(\alpha+\beta+1)} - 2 > 0 \tag{1.4.19}$$

where $r = \min_{1 \le i,j \le \tau}\{\alpha_i, \beta_j\} = \min\{\alpha, \beta\}$. Then,

(i) the inequality (1.1.3) has no eventually positive and nonincreasing solution;

(ii) the inequality (1.1.4) has no eventually negative and nondecreasing solution; and

(iii) the following equation has neither eventually positive and nonincreasing nor eventually negative and nondecreasing solutions

$$\Delta_m \Delta_n y(m,n) + \sum_{i=1}^{\tau} P_i(m,n,y(g_i(m),h_i(n))) = \sum_{i=1}^{\tau} Q_i(m,n,y(g_i(m),h_i(n))).$$

Proof. (i) Suppose that (1.1.3) has an eventually positive and nonincreasing solution $\{y(m,n)\}$. Let

$$V = \left\{ \lambda \in \mathbb{R} \;\middle|\; -\Delta_m \Delta_n y(m,n) + (1 - \lambda)y(m,n) + y(m+1, n+1) \le 0 \text{ eventually} \right\}.$$

We claim that V is nonempty. Indeed, from (1.1.3), (C1)–(C3) and (1.1.5), we find for sufficiently large m and n,

$$-\Delta_m \Delta_n y(m,n) \le -\sum_{i=1}^{\tau}[Q_i(m,n,y(g_i(m),h_i(n))) - P_i(m,n,y(g_i(m),h_i(n)))]$$

$$\le -\sum_{i=1}^{\tau} \mu_i(m,n) f_i(y(g_i(m),h_i(n)))$$

$$\le -\sum_{i=1}^{\tau} \gamma_i \mu_i(m,n) y(g_i(m),h_i(n))$$

$$\leq -\sum_{i=1}^{\tau} \gamma_i \mu_i(m,n) y(m-\alpha_i, n-\beta_i)$$

$$\leq -\sum_{i=1}^{\tau} \gamma_i \mu_i(m,n) y(m-\alpha, n-\beta). \qquad (1.4.20)$$

Using (1.4.20), we get

$$-\Delta_m \Delta_n y(m,n) + (1-2)y(m,n) + y(m+1, n+1)$$

$$\leq -\sum_{i=1}^{\tau} \gamma_i \mu_i(m,n) y(m-\alpha, n-\beta) - y(m,n) + y(m+1, n+1)$$

$$\leq -\sum_{i=1}^{\tau} \gamma_i \mu_i(m,n) y(m-\alpha, n-\beta) \leq 0.$$

Thus, $2 \in V$.

Further, if $\lambda \in V$, then it follows from the definition of V that (4.1) holds, and so it is necessary that $\lambda > 0$.

Next, since $y(m,n)$ is nonincreasing, we find that

$$-\Delta_m \Delta_n y(m,n) - y(m,n) + y(m+1, n+1)$$

$$\leq -\Delta_m \Delta_n y(m,n)$$

$$\leq -\sum_{i=1}^{\tau} \gamma_i \mu_i(m,n) y(m-\alpha, n-\beta)$$

$$\leq -\sum_{i=1}^{\tau} \rho_i \gamma_i \, 2^r \lambda^{-\alpha-\beta} y(m,n) \qquad (1.4.21)$$

where $\lambda \in V$ and we have applied Lemma 1.4.1 in the last inequality.

Now, consider the function $T(u) = u + u^{-\alpha-\beta} 2^r \sum_{i=1}^{\tau} \rho_i \gamma_i - 2$. It can easily be computed that

$$\min_{u>0} T(u) = T\left(\left[2^r (\alpha + \beta) \sum_{i=1}^{\tau} \rho_i \gamma_i \right]^{1/(\alpha+\beta+1)} \right) = d.$$

In particular, for $\lambda \in V$ we have $T(\lambda) \geq d$, which is equivalent to

$$\lambda^{-\alpha-\beta} 2^r \sum_{i=1}^{\tau} \rho_i \gamma_i \geq d + 2 - \lambda. \qquad (1.4.22)$$

Using (1.4.22) in (1.4.21), we find

$$-\Delta_m \Delta_n y(m,n) - y(m,n) + y(m+1, n+1) \leq -(d+2-\lambda)y(m,n)$$

or

$$-\Delta_m\Delta_n y(m,n) + [1 - (\lambda - d)]y(m,n) + y(m+1, n+1) \le 0$$

which implies that $(\lambda - d) \in V$.

By repeating the above procedure, we see that $(\lambda - id) \in V$, $i = 2, 3, 4, \cdots$. However, since $d > 0$, for sufficiently large i we have $\lambda - id < 0$. This contradicts the fact that elements of V are positive.

(ii) The proof is similar to that of (i).

(iii) This follows from (i) and (ii). □

Theorem 1.4.13. Suppose that $\alpha = \beta = 0$ and

$$\sum_{i=1}^{\tau} \gamma_i \mu_i(m,n) \ge 1 \qquad (1.4.23)$$

eventually. Then, the conclusion of Theorem 1.4.12 holds.

Proof. Again we shall only prove (i). For this, suppose that (1.1.3) has an eventually positive and nonincreasing solution $\{y(m,n)\}$. As before we have (1.4.20) which reduces to

$$-\Delta_m\Delta_n y(m,n) \le -\sum_{i=1}^{\tau} \gamma_i \mu_i(m,n) y(m,n)$$

or

$$y(m, n+1) + y(m+1, n) - y(m+1, n+1) \le \left[1 - \sum_{i=1}^{\tau} \gamma_i \mu_i(m,n)\right] y(m,n). \quad (1.4.24)$$

In view of (1.4.23), the right side of (1.4.24) is eventually nonpositive, whereas the left side is eventually positive due to the monotone nature of $y(m,n)$. □

Theorem 1.4.14. Suppose that

$$\limsup_{M,N\to\infty} \sum_{m=M}^{M+\alpha} \sum_{n=N}^{N+\beta} \sum_{i=1}^{\tau} \gamma_i \mu_i(m,n) > 1. \qquad (1.4.25)$$

Then, the conclusion of Theorem 1.4.12 holds.

Proof. Once again we shall only provide the proof of (i). Let $\{y(m,n)\}$ be an eventually positive and nonincreasing solution of (1.1.3). Then, we have (1.4.20) which on summing and using (1.4.4) gives

$$0 \ge \sum_{m=M}^{M+\alpha} \sum_{n=N}^{N+\beta} -\Delta_m\Delta_n y(m,n) + \sum_{m=M}^{M+\alpha} \sum_{n=N}^{N+\beta} \sum_{i=1}^{\tau} \gamma_i \mu_i(m,n) y(m-\alpha, n-\beta)$$

$$\geq y(M+\alpha+1,N) - y(M,N) + \sum_{m=M}^{M+\alpha}\sum_{n=N}^{N+\beta}\sum_{i=1}^{\tau}\gamma_i\mu_i(m,n)y(M+\alpha-\alpha,N+\beta-\beta)$$

$$= y(M+\alpha+1,N) + \left[\sum_{m=M}^{M+\alpha}\sum_{n=N}^{N+\beta}\sum_{i=1}^{\tau}\gamma_i\mu_i(m,n) - 1\right]y(M,N). \qquad (1.4.26)$$

In view of (1.4.25), the right side of (1.4.26) is positive. Hence, we get a contradiction. $\qquad\qquad\qquad\square$

Theorem 1.4.15. Let $\alpha, \beta > 0$. Suppose that there exists $b \in (0, 1/2]$ such that (1.4.10) holds for all large M and N. Further, assume that

$$\limsup_{M,N\to\infty}\sum_{m=M-\alpha}^{M}\sum_{n=N-\beta}^{N}\sum_{i=1}^{\tau}\gamma_i\mu_i(m,n) > c^2 - 1 \qquad (1.4.27)$$

where c is defined in Lemma 1.4.6. Then, the conclusion of Theorem 1.4.12 holds.

Proof. Once again suppose that (1.1.3) has an eventually positive and nonincreasing solution $\{y(m,n)\}$. As before, we have (1.4.20), a summation of which yields

$$\sum_{m=M-\alpha}^{M}\sum_{n=N-\beta}^{N} -\Delta_m\Delta_n y(m,n) \leq - \sum_{m=M-\alpha}^{M}\sum_{n=N-\beta}^{N}\sum_{i=1}^{\tau}\gamma_i\mu_i(m,n)y(m-\alpha,n-\beta)$$

$$\leq - \sum_{m=M-\alpha}^{M}\sum_{n=N-\beta}^{N}\sum_{i=1}^{\tau}\gamma_i\mu_i(m,n)y(M-\alpha,N-\beta). \qquad (1.4.28)$$

On the other hand, using (1.4.4) and Lemma 1.4.6 successively we find

$$\sum_{m=M-\alpha}^{M}\sum_{n=N-\beta}^{N} -\Delta_m\Delta_n y(m,n) \;\geq\; y(M+1,N-\beta) - y(M-\alpha,N-\beta)$$

$$\geq\; y(M+1,N+1) - y(M-\alpha,N-\beta)$$

$$\geq\; (1-c^2)y(M+1,N+1). \qquad (1.4.29)$$

Coupling (1.4.28) and (1.4.29), we get

$$\sum_{m=M-\alpha}^{M}\sum_{n=N-\beta}^{N}\sum_{i=1}^{\tau}\gamma_i\mu_i(m,n)y(M-\alpha,N-\beta) \;\leq\; (c^2-1)y(M+1,N+1)$$

$$\leq\; (c^2-1)y(M-\alpha,N-\beta)$$

or equivalently

$$\sum_{m=M-\alpha}^{M} \sum_{n=N-\beta}^{N} \sum_{i=1}^{\tau} \gamma_i \mu_i(m,n) \leq c^2 - 1$$

which is a contradiction to (1.4.27). □

Theorem 1.4.16. Let $\alpha, \beta > 0$. Suppose that there exists $b \in (0,2]$ such that (1.4.10) holds for all large M and N. Further, assume that

$$\limsup_{M,N\to\infty} \sum_{m=M-\alpha}^{M} \sum_{n=N-\beta}^{N} \sum_{i=1}^{\tau} \gamma_i \mu_i(m,n) > \frac{16}{b^4} - 1. \qquad (1.4.30)$$

Then, the conclusion of Theorem 1.4.12 holds.

Proof. The proof is similar to that of Theorem 1.4.15 with the modification that we apply Lemma 1.4.7 instead of Lemma 1.4.6. □

Theorem 1.4.17. Let $\beta = 0$, $\alpha > 0$. Suppose that there exists $b \in (0, 1/2]$ such that (1.4.13) holds for all large M and N. Further, assume that

$$\limsup_{M,N\to\infty} \sum_{m=M-\alpha}^{M} \sum_{i=1}^{\tau} \gamma_i \mu_i(m,N) > 1 - \frac{1}{c} \qquad (1.4.31)$$

where c is defined in Lemma 1.4.6. Then, the conclusion of Theorem 1.3.2 holds.

Proof. Let $\{y(m,n)\}$ be an eventually positive and nonincreasing solution of (1.1.3). Then, it follows from (1.4.20) that

$$-\sum_{i=1}^{\tau} \gamma_i \mu_i(m,n) y(m-\alpha,n)$$

$$\geq -\Delta_m \Delta_n y(m,n)$$

$$= y(m+1,n) + y(m,n+1) - y(m+1,n+1) - y(m,n)$$

$$\geq y(m+1,n) - y(m,n). \qquad (1.4.32)$$

Using the above inequality, we find for large M and N,

$$y(M+1,N) - y(M-\alpha,N) = \sum_{m=M-\alpha}^{M} [y(m+1,N) - y(m,N)]$$

$$\leq -\sum_{m=M-\alpha}^{M} \sum_{i=1}^{\tau} \gamma_i \mu_i(m,N) y(m-\alpha,N)$$

$$\leq -\sum_{m=M-\alpha}^{M} \sum_{i=1}^{\tau} \gamma_i \mu_i(m,N) y(M-\alpha,N)$$

which is the same as

$$y(M+1, N) \leq \left[1 - \sum_{m=M-\alpha}^{M} \sum_{i=1}^{\tau} \gamma_i \mu_i(m, N) \right] y(M - \alpha, N). \qquad (1.4.33)$$

On the other hand, by means of Lemma 1.4.8 we have the following for large M and N,

$$y(M+1, N) \geq \frac{1}{c} \, y(M - \alpha, N)$$

which can be applied in (1.4.33) to give

$$\left[1 - \sum_{m=M-\alpha}^{M} \sum_{i=1}^{\tau} \gamma_i \mu_i(m, N) - \frac{1}{c} \right] y(M - \alpha, N) \geq 0.$$

Hence, we obtain

$$1 - \sum_{m=M-\alpha}^{M} \sum_{i=1}^{\tau} \gamma_i \mu_i(m, N) - \frac{1}{c} \geq 0$$

for all large M and N, which is a contradiction to (1.4.31). $\qquad \square$

Theorem 1.4.18. Let $\alpha = 0$, $\beta > 0$. Suppose that there exists $b \in (0, 1/2]$ such that (1.4.15) holds for all large M and N. Further, assume that

$$\limsup_{M, N \to \infty} \sum_{n=N-\beta}^{N} \sum_{i=1}^{\tau} \gamma_i \mu_i(M, n) > 1 - \frac{1}{c} \qquad (1.4.34)$$

where c is defined in Lemma 1.4.6. Then, the conclusion of Theorem 1.3.2 holds.

Proof. Suppose that $\{y(m, n)\}$ is an eventually positive and nonincreasing solution of (1.1.3). Then, it is clear from (1.4.20) that

$$-\sum_{i=1}^{\tau} \gamma_i \mu_i(m, n) y(m, n - \beta) \geq -\Delta_m \Delta_n y(m, n) \geq y(m, n+1) - y(m, n).$$

Applying the above inequality, we find for large M and N,

$$\begin{aligned}
y(M, N+1) - y(M, N - \beta) &= \sum_{n=N-\beta}^{N} [y(M, n+1) - y(M, n)] \\
&\leq -\sum_{n=N-\beta}^{N} \sum_{i=1}^{\tau} \gamma_i \mu_i(M, n) y(M, n - \beta) \\
&\leq -\sum_{n=N-\beta}^{N} \sum_{i=1}^{\tau} \gamma_i \mu_i(M, n) y(M, N - \beta)
\end{aligned}$$

which, in view of Lemma 1.4.9, further leads to

$$\left[1 - \sum_{n=N-\beta}^{N} \sum_{i=1}^{\tau} \gamma_i \mu_i(M, n) - \frac{1}{c} \right] y(M, N - \beta) \geq 0.$$

Clearly, this is not possible in view of (1.4.34). □

Theorem 1.4.19. Let $\beta = 0$, $\alpha > 0$. Suppose that there exists $b > 0$ such that (1.4.13) holds for all large M and N. Further, assume that

$$\limsup_{M,N \to \infty} \sum_{m=M-\alpha}^{M} \sum_{i=1}^{\tau} \gamma_i \mu_i(m, N) > 1 - \frac{b^2}{4}. \qquad (1.4.35)$$

Then, the conclusion of Theorem 1.4.12 holds.

Proof. The proof is similar to that of Theorem 1.4.17 with the modification that Lemma 1.4.10 is used instead of Lemma 1.4.8. □

Theorem 1.4.20. Let $\alpha = 0$, $\beta > 0$. Suppose that there exists $b > 0$ such that (1.4.15) holds for all large M and N. Further, assume that

$$\limsup_{M,N \to \infty} \sum_{n=N-\beta}^{N} \sum_{i=1}^{\tau} \gamma_i \mu_i(M, n) > 1 - \frac{b^2}{4}. \qquad (1.4.36)$$

Then, the conclusion of Theorem 1.4.12 holds.

Proof. The proof is similar to that of Theorem 1.4.18 with the modification that Lemma 1.4.11 is used instead of Lemma 1.4.9. □

Theorem 1.4.21. Let $\alpha, \beta > 0$. Suppose that

$$\liminf_{M,N \to \infty} \frac{1}{\alpha\beta} \sum_{m=M-\alpha}^{M-1} \sum_{n=N-\beta}^{N-1} \sum_{i=1}^{\tau} \gamma_i \mu_i(m, n) > w \qquad (1.4.37)$$

where $w = 2(2\nu)^\nu (\nu + 1)^{-\nu-1}$ and $\nu = 2\alpha\beta(\alpha + \beta)^{-1}$. Then, the conclusion of Theorem 1.4.12 holds.

Proof. Once again suppose that (1.1.3) has an eventually positive and nonincreasing solution $\{y(m, n)\}$. As before, we have (1.4.20) from which it follows that

$$y(m + 1, n) + y(m, n + 1) - y(m + 1, n + 1) - y(m, n)$$

$$\leq -\sum_{i=1}^{\tau} \gamma_i \mu_i(m, n) y(m - \alpha, n - \beta) \leq -\sum_{i=1}^{\tau} \gamma_i \mu_i(m, n) y(m, n). \qquad (1.4.38)$$

Subsequently, on dividing (1.4.38) by $y(m, n)$ and noting that $\dfrac{y(m+1, n+1)}{y(m, n)} \leq$ 1, we find

$$\sum_{i=1}^{\tau} \gamma_i \mu_i(m, n) \leq 2 - \frac{y(m+1, n) + y(m, n+1)}{y(m, n)}$$

$$\leq 2 - \frac{[y(m+1, n) y(m, n+1)]^{1/2}}{y(m, n)}. \qquad (1.4.39)$$

Next, in view of (1.4.37) we may choose $r > 0$ such that for all large M and N,

$$\frac{1}{\alpha \beta} \sum_{m=M-\alpha}^{M-1} \sum_{n=N-\beta}^{N-1} \sum_{i=1}^{\tau} \gamma_i \mu_i(m, n) \geq r > w. \qquad (1.4.40)$$

Using (1.4.39) it is immediate that

$$\frac{1}{\alpha \beta} \sum_{m=M-\alpha}^{M-1} \sum_{n=N-\beta}^{N-1} \left\{ 2 - \frac{[y(m+1, n) y(m, n+1)]^{1/2}}{y(m, n)} \right\} \geq r$$

or

$$2 - \frac{1}{\alpha \beta} \sum_{m=M-\alpha}^{M-1} \sum_{n=N-\beta}^{N-1} \frac{[y(m+1, n) y(m, n+1)]^{1/2}}{y(m, n)} \geq r. \qquad (1.4.41)$$

Now, we shall show that

$$\frac{1}{\alpha \beta} \sum_{m=M-\alpha}^{M-1} \sum_{n=N-\beta}^{N-1} \frac{[y(m+1, n) y(m, n+1)]^{1/2}}{y(m, n)} \geq \left[\frac{y(M, N)}{y(M-\alpha, N-\beta)} \right]^{1/\nu}. \qquad (1.4.42)$$

In fact, by applying the arithmetic-geometric mean inequality we find

$$\frac{1}{\alpha \beta} \sum_{m=M-\alpha}^{M-1} \sum_{n=N-\beta}^{N-1} \frac{[y(m+1, n) y(m, n+1)]^{1/2}}{y(m, n)}$$

$$\geq \frac{1}{\alpha} \sum_{m=M-\alpha}^{M-1} \left\{ \prod_{n=N-\beta}^{N-1} \frac{[y(m+1, n) y(m, n+1)]^{1/2}}{y(m, n)} \right\}^{1/\beta}$$

$$\geq \left\{ \prod_{m=M-\alpha}^{M-1} \prod_{n=N-\beta}^{N-1} \frac{[y(m+1, n) y(m, n+1)]^{1/2}}{y(m, n)} \right\}^{1/(\alpha\beta)}$$

$$= \prod_{m=M-\alpha}^{M-1} \left[\prod_{n=N-\beta}^{N-1} \frac{y(m, n+1)}{y(m, n)} \frac{y(m+1, n)}{y(m, n)} \right]^{1/(2\alpha\beta)}$$

$$= \prod_{m=M-\alpha}^{M-1} \left[\frac{y(m,N)}{y(m,N-\beta)} \prod_{n=N-\beta}^{N-1} \frac{y(m+1,n)}{y(m,n)} \right]^{1/(2\alpha\beta)}$$

$$= \left\{ \left[\prod_{m=M-\alpha}^{M-1} \frac{y(m,N)}{y(m,N-\beta)} \right] \left[\prod_{n=N-\beta}^{N-1} \prod_{m=M-\alpha}^{M-1} \frac{y(m+1,n)}{y(m,n)} \right] \right\}^{1/(2\alpha\beta)}$$

$$= \left\{ \left[\prod_{m=M-\alpha}^{M-1} \frac{y(m,N)}{y(m,N-\beta)} \right] \left[\prod_{n=N-\beta}^{N-1} \frac{y(M,n)}{y(M-\alpha,n)} \right] \right\}^{1/(2\alpha\beta)}$$

$$\geq \left\{ \left[\prod_{m=M-\alpha}^{M-1} \frac{y(M,N)}{y(M-\alpha,N-\beta)} \right] \left[\prod_{n=N-\beta}^{N-1} \frac{y(M,N)}{y(M-\alpha,N-\beta)} \right] \right\}^{1/(2\alpha\beta)}$$

$$= \left[\frac{y(M,N)}{y(M-\alpha,N-\beta)} \right]^{1/\nu}.$$

Substituting (1.4.42) into (1.4.41), we get

$$2 - \left[\frac{y(M,N)}{y(M-\alpha,N-\beta)} \right]^{1/\nu} \geq r$$

or

$$2 - r \geq \left[\frac{y(M,N)}{y(M-\alpha,N-\beta)} \right]^{1/\nu} > 0. \qquad (1.4.43)$$

Thus, it follows that $r \in (0,2)$.

For the function $T(u) = (2-u)u^{1/\nu}$, it can be easily computed that

$$\max_{0 \leq u \leq 2} T(u) = T\left(\frac{2}{\nu+1} \right) = w^{1/\nu}.$$

In particular, we have $T(r) \leq w^{1/\nu}$ which is equivalent to

$$2 - r \leq \left(\frac{w}{r} \right)^{1/\nu}. \qquad (1.4.44)$$

Using (1.4.44) in (1.4.43) provides

$$\frac{y(M-\alpha,N-\beta)}{y(M,N)} \geq \frac{r}{w}. \qquad (1.4.45)$$

Now, if we use (1.4.45) in (1.4.20), we get the following inequality which corresponds to (1.4.38)

$$y(m+1,n) + y(m,n+1) - y(m+1,n+1) - y(m,n)$$

$$\leq -\sum_{i=1}^{\tau} \gamma_i \mu_i(m,n) y(m-\alpha,n-\beta) \leq -\sum_{i=1}^{\tau} \gamma_i \mu_i(m,n) \frac{r}{w} \, y(m,n). \qquad (1.4.38)'$$

By a similar argument, we obtain the following corresponding relations

$$\frac{r}{w} \sum_{i=1}^{\tau} \gamma_i \mu_i(m,n) \leq 2 - \frac{[y(m+1,n)y(m,n+1)]^{1/2}}{y(m,n)}, \qquad (1.4.39)'$$

$$2 - \frac{1}{\alpha\beta} \sum_{m=M-\alpha}^{M-1} \sum_{n=N-\beta}^{N-1} \frac{[y(m+1,n)y(m,n+1)]^{1/2}}{y(m,n)} \geq \frac{r^2}{w}, \qquad (1.4.41)'$$

$$2 - \frac{r^2}{w} \geq \left[\frac{y(M,N)}{y(M-\alpha,N-\beta)} \right]^{1/\nu} > 0, \qquad (1.4.43)'$$

$$2 - \frac{r^2}{w} \leq \left(\frac{w^2}{r^2} \right)^{1/\nu}, \qquad (1.4.44)'$$

$$\frac{y(M-\alpha,N-\beta)}{y(M,N)} \geq \left(\frac{r}{w} \right)^2. \qquad (1.4.45)'$$

By repeating the procedure, we find that

$$\frac{y(M-\alpha,N-\beta)}{y(M,N)} \geq \left(\frac{r}{w} \right)^{\ell} \qquad (1.4.46)$$

where $\ell = 1,2,\cdots$. Since $r/w > 1$, it follows that

$$\frac{y(M-\alpha,N-\beta)}{y(M,N)} \to \infty$$

as $\ell \to \infty$. However, from Lemma 1.4.7 we have (1.4.12) which leads to

$$\frac{y(M-\alpha,N-\beta)}{y(M,N)} \leq \frac{y(M-\alpha,N-\beta)}{y(M+1,N+1)} \leq \frac{16}{b^4}.$$

Hence, we get a contradiction. $\qquad\qquad\qquad\qquad\qquad\qquad\qquad\qquad \square$

Theorem 1.4.22. Let $\beta = 0$, $\alpha > 0$. Suppose that

$$\liminf_{M,N\to\infty} \frac{1}{\alpha} \sum_{m=M-\alpha}^{M-1} \sum_{i=1}^{\tau} \gamma_i \mu_i(m,N) > w_1 \qquad (1.4.47)$$

where $w_1 = \alpha^\alpha (\alpha+1)^{-\alpha-1}$. Then, the conclusion of Theorem 1.4.12 holds.

Proof. Let $\{y(m,n)\}$ be an eventually positive and nonincreasing solution of (1.1.3). Then, we have (1.4.32) which provides

$$y(m+1,n) - y(m,n) \leq -\sum_{i=1}^{\tau} \gamma_i \mu_i(m,n) y(m-\alpha,n) \leq -\sum_{i=1}^{\tau} \gamma_i \mu_i(m,n) y(m,n)$$

$$(1.4.48)$$

or

$$\sum_{i=1}^{\tau} \gamma_i \mu_i(m,n) \leq 1 - \frac{y(m+1,n)}{y(m,n)}. \qquad (1.4.49)$$

Next, condition (1.4.47) implies the existence of some $r > 0$ such that for all large M and N,

$$\frac{1}{\alpha} \sum_{m=M-\alpha}^{M-1} \sum_{i=1}^{\tau} \gamma_i \mu_i(m,N) \geq r > w_1. \qquad (1.4.50)$$

Using (1.4.49) in (1.4.50) yields

$$r \leq \frac{1}{\alpha} \sum_{m=M-\alpha}^{M-1} \left[1 - \frac{y(m+1,N)}{y(m,N)} \right]$$

$$\leq 1 - \left[\prod_{m=M-\alpha}^{M-1} \frac{y(m+1,N)}{y(m,N)} \right]^{1/\alpha} = 1 - \left[\frac{y(M,N)}{y(M-\alpha,N)} \right]^{1/\alpha}. \qquad (1.4.51)$$

Therefore, we find that $r \in (0,1)$.

For the function $T(u) = (1-u)u^{1/\alpha}$, it is clear that

$$\max_{0 \leq u \leq 1} T(u) = T\left(\frac{1}{\alpha+1} \right) = w_1^{1/\alpha}.$$

In particular, we have $T(r) \leq w_1^{1/\alpha}$ or

$$1 - r \leq \left(\frac{w_1}{r} \right)^{1/\alpha}$$

which we use in (1.4.51) to get

$$\frac{y(M-\alpha,N)}{y(M,N)} \geq \frac{r}{w_1}. \qquad (1.4.52)$$

Now, applying (1.4.52) in (1.4.32) and then following a similar argument, we obtain the following corresponding relations:

$$y(m+1,n) - y(m,n) \leq -\sum_{i=1}^{\tau} \gamma_i \mu_i(m,n) y(m-\alpha,n) \leq -\sum_{i=1}^{\tau} \gamma_i \mu_i(m,n) \frac{r}{w_1} y(m,n),$$

$$(1.4.48)'$$

$$\frac{r}{w_1} \sum_{i=1}^{\tau} \gamma_i \mu_i(m,n) \le 1 - \frac{y(m+1,n)}{y(m,n)}, \tag{1.4.49$'$}$$

$$\frac{r^2}{w_1} \le 1 - \left[\frac{y(M,N)}{y(M-\alpha,N)} \right]^{1/\alpha}, \tag{1.4.51$'$}$$

$$\frac{y(M-\alpha,N)}{y(M,N)} \ge \left(\frac{r}{w_1} \right)^2. \tag{1.4.52$'$}$$

Hence, by repeating the procedure we find that

$$\frac{y(M-\alpha,N)}{y(M,N)} \ge \left(\frac{r}{w_1} \right)^{\ell}$$

where $\ell = 1, 2, \cdots$. Since $r/w_1 > 1$, it follows that

$$\frac{y(M-\alpha,N)}{y(M,N)} \to \infty \tag{1.4.53}$$

as $\ell \to \infty$. However, from Lemma 1.4.10 we have

$$\frac{y(M-\alpha,N)}{y(M,N)} \le \frac{y(M-\alpha,N)}{y(M+1,N)} \le \frac{4}{b^2}$$

which contradicts (1.4.53). $\qquad\qquad\qquad\qquad\qquad\qquad\qquad\qquad\square$

Theorem 1.4.23. Let $\alpha = 0$, $\beta > 0$. Suppose that

$$\liminf_{M,N \to \infty} \frac{1}{\beta} \sum_{n=N-\beta}^{N-1} \sum_{i=1}^{\tau} \gamma_i \mu_i(M,n) > w_2 \tag{1.4.54}$$

where $w_2 = \beta^{\beta}(\beta+1)^{-\beta-1}$. Then, the conclusion of Theorem 1.4.12 holds.

Proof. The proof is similar to that of Theorem 1.4.22 with the modification that Lemma 1.4.11 is used instead of Lemma 1.4.10. $\qquad\qquad\qquad\qquad\qquad\square$

Theorem 1.4.24. Let $\alpha, \beta > 0$. Suppose that

$$\liminf_{m,n \to \infty} \gamma_i \mu_i(m,n) - 1 = v_i > 0 \tag{1.4.55}$$

and

$$\sum_{i=1}^{\tau} \left[\limsup_{m,n \to \infty} \gamma_i \mu_i(m,n) + 2v_i \right] > 4 - 2\tau. \tag{1.4.56}$$

Then, the conclusion of Theorem 1.4.12 holds.

Proof. Let $\{y(m,n)\}$ be an eventually positive and nonincreasing solution of (1.1.3). Then, applying (1.4.20) we find

$$-\Delta_m\Delta_n y(m,n)+y(m+1,n+1) \leq -\sum_{i=1}^{\tau}\gamma_i\mu_i(m,n)y(m-\alpha,n-\beta)+y(m-\alpha,n-\beta)$$

which is equivalent to

$$y(m+1,n)+y(m,n+1) \leq y(m,n)+\left[1-\sum_{i=1}^{\tau}\gamma_i\mu_i(m,n)\right]y(m-\alpha,n-\beta). \quad (1.4.57)$$

Since the left side of (1.4.57) is eventually positive, it is necessary that for all large m and n,

$$y(m,n) \geq \left[\sum_{i=1}^{\tau}\gamma_i\mu_i(m,n)-1\right]y(m-\alpha,n-\beta)$$

$$\geq \left[\sum_{i=1}^{\tau}\gamma_i\mu_i(m,n)-1\right]y(m-1,n-1)$$

which in view of (1.4.55) provides

$$y(m,n) > \left\{\sum_{i=1}^{\tau}[(v_i-\epsilon)+1]-1\right\}y(m-1,n-1)$$

$$= \left[\sum_{i=1}^{\tau}(v_i-\epsilon)+\tau-1\right]y(m-1,n-1) \quad (1.4.58)$$

where $\epsilon \in (0,\min_{1\leq i\leq\tau}v_i)$ is arbitrarily small.

It is clear from (1.4.58) that

$$y(m+1,n) > \left[\sum_{i=1}^{\tau}(v_i-\epsilon)+\tau-1\right]y(m,n-1) \geq \left[\sum_{i=1}^{\tau}(v_i-\epsilon)+\tau-1\right]y(m,n)$$

$$(1.4.59)$$

and

$$y(m,n+1) > \left[\sum_{i=1}^{\tau}(v_i-\epsilon)+\tau-1\right]y(m-1,n) \geq \left[\sum_{i=1}^{\tau}(v_i-\epsilon)+\tau-1\right]y(m,n).$$

$$(1.4.60)$$

Now, using (1.4.20), (1.4.59) and (1.4.60) it follows from (1.1.3) that

$$0 \geq \sum_{i=1}^{\tau}[Q_i(m,n,y(g_i(m),h_i(n)))-P_i(m,n,y(g_i(m),h_i(n)))]-\Delta_m\Delta_n y(m,n)$$

$$\geq \sum_{i=1}^{\tau} \gamma_i \mu_i(m,n) y(m-\alpha, n-\beta) + y(m+1,n) + y(m,n+1) - y(m,n) - y(m+1,n+1)$$

$$\geq \sum_{i=1}^{\tau} \gamma_i \mu_i(m,n) y(m,n) + y(m+1,n) + y(m,n+1) - y(m,n) - y(m,n)$$

$$> \left\{ \sum_{i=1}^{\tau} \gamma_i \mu_i(m,n) + 2 \left[\sum_{i=1}^{\tau} (v_i - \epsilon) + \tau - 1 \right] - 2 \right\} y(m,n)$$

$$= \left[\sum_{i=1}^{\tau} \gamma_i \mu_i(m,n) + 2 \sum_{i=1}^{\tau} (v_i - \epsilon) + 2\tau - 4 \right] y(m,n).$$

Therefore, we must have

$$\sum_{i=1}^{\tau} \gamma_i \mu_i(m,n) + 2 \sum_{i=1}^{\tau} (v_i - \epsilon) + 2\tau - 4 < 0$$

which is a contradiction to (1.4.56). $\qquad\qquad\qquad\qquad\qquad\qquad\square$

Remark 1.4.25. Let $\tau \geq 2$. Then, with condition (1.4.55) given, the inequality (1.4.56) is automatically satisfied (left side is positive, right side is nonpositive). Hence, if $\tau \geq 2$ and (1.4.55) holds, then the conclusion of Theorem 1.4.12 follows.

Theorem 1.4.26. Suppose that

$$\liminf_{m,n \to \infty} \mu_i(m,n) = \mu_i > 0 \qquad\qquad\qquad (1.4.61)$$

and

$$\sum_{i=1}^{\tau} \gamma_i \mu_i \frac{(r_i+1)^{r_i+1}}{r_i^{r_i}} > 1 \qquad\qquad\qquad (1.4.62)$$

where $r_i = \min\{\alpha_i, \beta_i\}$, $1 \leq i \leq \tau$ and $0^0 = 1$. Then, the conclusion of Theorem 1.4.12 holds.

Proof. Suppose that (1.1.3) has an eventually positive and nonincreasing solution $\{y(m,n)\}$. Define

$$z(m,n) = \frac{y(m,n)}{y(m+1,n+1)} \quad (\geq 1).$$

Then, it is clear that

$$\frac{y(m,n+1) + y(m+1,n)}{y(m,n)} \geq \frac{y(m+1,n+1) + y(m+1,n+1)}{y(m,n)} = \frac{2}{z(m,n)}.$$

$$(1.4.63)$$

Now, dividing (1.4.20) by $y(m, n)$ yields

$$\frac{y(m, n+1) + y(m, n+1)}{y(m, n)} - \frac{1}{z(m, n)} - 1 \leq -\sum_{i=1}^{\tau} \gamma_i \mu_i(m, n) \frac{y(m - \alpha_i, n - \beta_i)}{y(m, n)}$$

$$\leq -\sum_{i=1}^{\tau} \gamma_i \mu_i(m, n) \frac{y(m - r_i, n - r_i)}{y(m, n)}$$

$$= -\sum_{i=1}^{\tau} \gamma_i \mu_i(m, n) \prod_{j=1}^{r_i} z(m - j, n - j).$$

In view of (1.4.63), the above inequality gives rise to

$$\frac{1}{z(m, n)} - 1 \leq -\sum_{i=1}^{\tau} \gamma_i \mu_i(m, n) \prod_{j=1}^{r_i} z(m - j, n - j). \tag{1.4.64}$$

Note that $r_i \neq 0$ for some $1 \leq i \leq \tau$, since otherwise (1.4.64) provides

$$-1 < \frac{1}{z(m, n)} - 1 \leq -\sum_{i=1}^{\tau} \gamma_i \mu_i(m, n)$$

and so

$$-1 \leq -\sum_{i=1}^{\tau} \gamma_i \mu_i,$$

whereas in this case (1.4.62) is the same as

$$\sum_{i=1}^{\tau} \gamma_i \mu_i > 1.$$

Further, $z(m, n) < \infty$, for otherwise from (1.4.64) we get $-1 \leq -\infty$.

Denoting $z = \liminf_{m,n \to \infty} z(m, n) \in (1, \infty)$, it follows from (1.4.64) that

$$\frac{1}{z} - 1 \leq -\sum_{i=1}^{\tau} \gamma_i \mu_i z^{r_i}$$

or

$$\sum_{i=1}^{\tau} \gamma_i \mu_i \frac{z^{r_i + 1}}{z - 1} \leq 1. \tag{1.4.65}$$

Now, for the function $T(u) = u^{r_i + 1}(u - 1)^{-1}$, it can be verified that

$$\min_{u > 1} T(u) = T\left(\frac{r_i + 1}{r_i}\right) = \frac{(r_i + 1)^{r_i + 1}}{r_i^{r_i}}. \tag{1.4.66}$$

Thus, (1.4.65) implies that

$$\sum_{i=1}^{\tau} \gamma_i \mu_i \frac{(r_i + 1)^{r_i+1}}{r_i^{r_i}} \leq 1$$

which is a contradiction to (1.4.62). $\qquad \square$

Theorem 1.4.27. Suppose that (1.4.61) holds and

$$\tau \left(\prod_{i=1}^{\tau} \gamma_i \mu_i \right)^{1/\tau} > \frac{\eta^\eta}{(\eta + 1)^{\eta+1}} \qquad (1.4.67)$$

where $\eta = \sum_{i=1}^{\tau} r_i / \tau$ and $r_i = \min\{\alpha_i, \beta_i\}$, $1 \leq i \leq \tau$. Then, the conclusion of Theorem 1.4.12 holds.

Proof. Again let $\{y(m, n)\}$ be an eventually positive and nonincreasing solution of (1.1.3). As in the proof of Theorem 1.4.26, we still get (1.4.65) which, upon using the well known inequality between arithmetic and geometric means, leads to

$$1 \geq \sum_{i=1}^{\tau} \gamma_i \mu_i \frac{z^{r_i+1}}{z-1} \geq \frac{\tau}{z-1} \left(\prod_{i=1}^{\tau} \gamma_i \mu_i z^{r_i+1} \right)^{1/\tau} = \frac{\tau}{z-1} z^{\eta+1} \left(\prod_{i=1}^{\tau} \gamma_i \mu_i \right)^{1/\tau}.$$

In view of (1.4.66), the above inequality implies

$$1 \geq \tau \frac{(\eta+1)^{\eta+1}}{\eta^\eta} \left(\prod_{i=1}^{\tau} \gamma_i \mu_i \right)^{1/\tau}$$

which contradicts (1.4.67). $\qquad \square$

Theorem 1.4.28. Suppose that (1.4.61) holds and

$$\sum_{i=1}^{\tau} \gamma_i \mu_i > \frac{r^r}{(r + 1)^{r+1}} \qquad (1.4.68)$$

where $r = \min_{1 \leq i,j \leq \tau}\{\alpha_i, \beta_j\} = \min\{\alpha, \beta\}$. Then, the conclusion of Theorem 1.4.12 holds.

Proof. Once again suppose that (1.1.3) has an eventually positive and nonincreasing solution $\{y(m, n)\}$. As in the proof of Theorem 1.4.26, we have (1.4.65). Since $z > 1$, it is clear from (1.4.65) that

$$\sum_{i=1}^{\tau} \gamma_i \mu_i \frac{z^{r+1}}{z-1} \leq 1$$

which in view of (1.4.66) provides

$$\sum_{i=1}^{\tau} \gamma_i \mu_i \frac{(r+1)^{r+1}}{r^r} \leq 1.$$

Hence, we get a contradiction to (1.4.68).　　　　　　　　　　　□

1.5　Examples

The examples in this section dwell upon the importance of the results obtained in Sections 2–4. Not only that we demonstrate situations when the conditions of the theorems are satisfied, then the partial difference equation considered satisfies the conclusions of the theorems, we also include examples which show that if the conditions of the theorems are violated, then the partial difference equation considered do not fulfill the conclusions of the theorems.

Example 1.5.1. Consider the partial difference equation

$$\nabla_m \nabla_n y(m,n) = \frac{2(2n-1)(n+3)}{n(n-1)} y(m+1,n+3), \ m \geq 1, \ n \geq 2. \quad (1.5.1)$$

Here, $k = 1$, $\ell = 3$, $\alpha = 1$ and $\beta = 3$. Choosing $f(u) = u$, we have $\gamma = 1$. Further, since

$$\frac{Q(m,n,y(m+1,n+3))}{f(y(m+1,n+3))} = \frac{2(2n-1)(n+3)}{n(n-1)},$$

$$\frac{P(m,n,y(m+1,n+3))}{f(y(m+1,n+3))} = 0,$$

we may take

$$q(m,n) = q'(m,n) = \frac{2(2n-1)(n+3)}{n(n-1)}, \quad p(m,n) = p'(m,n) = 0.$$

Thus, (A1) - (A3) are fulfilled.

Case 1: Corollary 1.2.2

The left side of (1.2.19) is

$$\liminf_{m,n\to\infty} \frac{1}{3} \sum_{i=1}^{1} \sum_{j=1}^{3} \left[\frac{2(2(n+j)-1)(n+j+3)}{(n+j)(n+j-1)} - 1 \right] = \frac{1}{3} \cdot 9 = 3$$

which is more than the right side $(= 27/256)$.

Case 2: Corollary 1.2.4

We find that

$$\liminf_{m,n\to\infty} \mu(m,n) = 4 > \frac{1}{\gamma}\left[\frac{\beta^\beta}{2^\alpha(1+\beta)^{1+\beta}}+1\right] = \frac{3^3}{2\cdot 4^4}+1$$

and so (1.2.30) is satisfied.

Case 3: Corollary 1.2.6

We have

$$\mu(m,n) = \frac{2(2n-1)(n+3)}{n(n-1)} = 2\left(2-\frac{1}{n}\right)\left(1+\frac{4}{n-1}\right) \ge 2\left(2-\frac{1}{2}\right)\cdot 1 = 3 \equiv c$$

and

$$\frac{1}{\gamma}\left[\frac{(k+\ell)^{k+\ell}}{(k+\ell+1)^{k+\ell+1}}+1\right] = 1.08192.$$

Hence, (1.2.36) is fulfilled.

Case 4: Theorem 1.3.2

Here, $\tau = 1$, $q_1 = 4$ and $p_1' = 0$. The left side of (1.3.2) is 16 which is more than 1.

Case 5: Theorem 1.3.3

We see that (1.3.9) holds as

$$\limsup_{m,n\to\infty} \sum_{i=m-1}^{m}\sum_{j=n-3}^{n}\frac{2(2j-1)(j+3)}{j(j-1)}$$

$$= 2\limsup_{n\to\infty}\sum_{j=n-3}^{n}\frac{2(2j-1)(j+3)}{j(j-1)} = 32 > \ell' + 2 = 5.$$

Hence, it follows from Corollaries 1.2.2, 1.2.4, 1.2.6, Theorems 1.3.2 and 1.3.3 that equation (1.5.1) has no unbounded nonoscillatory solution. In fact, (1.5.1) has oscillatory solutions given by $\{y(m,n)\} = \left\{(-1)^{m\pm n}\frac{1}{n}\right\}$.

Example 1.5.2. Consider the partial difference equation

$$\nabla_m\nabla_n y(m,n) = \frac{(2m-1)(m+2)(2n-1)(n+4)}{m(m-1)n(n-1)} \, y(m+2,n+4), \; m,n \ge 2.$$

$$(1.5.2)$$

Taking $f(u) = u$, we have $\gamma = 1$. Further, we may choose

$$q(m,n) = q'(m,n) = \frac{(2m-1)(m+2)(2n-1)(n+4)}{m(m-1)n(n-1)}, \; p(m,n) = p'(m,n) = 0.$$

Clearly, (A1) - (A3) are satisfied. It can be checked that all the conditions of Corollaries 1.2.2, 1.2.4, 1.2.6, Theorems 1.3.2 and 1.3.3 are fulfilled. Therefore, we conclude that (1.5.2) has no unbounded nonoscillatory solution. In fact, (1.5.2) has oscillatory solutions given by $\{y(m,n)\} = \left\{(-1)^{m\pm n}\dfrac{1}{mn}\right\}$.

Example 1.5.3. Consider the partial difference equation

$$\nabla_m \nabla_n y(m,n) = \frac{2(n+\ell_1)}{n} \, y(m+k_1, n+\ell_1) +$$

$$\frac{2(n+\ell_2)}{n-1} \, y(m+k_2, n+\ell_2), m \geq 1, n \geq 2 \quad (1.5.3)$$

where $(k_1 + \ell_1)$ and $(k_2 + \ell_2)$ are even.

Here, $\tau = 2$. Choosing $f_1(u) = f_2(u) = u$, we have $\gamma_1 = \gamma_2 = 1$. Let

$$q_1(m,n) = q_1'(m,n) = \frac{2(n+\ell_1)}{n}, \quad q_2(m,n) = q_2'(m,n) = \frac{2(n+\ell_2)}{n-1},$$

$$p_i(m,n) = p_i'(m,n) = 0, \; i = 1, 2.$$

Then, we have $q_i - p_i' = 2, \; i = 1, 2$.

It is now obvious that the right side of (1.3.2) is more than 1. Condition (1.3.9) also holds as

$$\limsup_{m,n\to\infty} \sum_{s=1}^{2} \gamma_s \sum_{i=m-k'}^{m} \sum_{j=n-\ell'}^{n} [q_s(i,j) - p_s'(i,j)]$$

$$= \limsup_{m,n\to\infty} \sum_{i=m-k'}^{m} \sum_{j=n-\ell'}^{n} \left[\frac{2(j+\ell_1)}{j} + \frac{2(j+\ell_2)}{j-1}\right]$$

$$= 4(k'+1)(\ell'+1) > \ell'+2.$$

Hence, by Theorems 1.3.2 and 1.3.3 equation (1.5.3) has no unbounded nonoscillatory solution. In fact, (1.5.3) has oscillatory solutions given by $\{y(m,n)\} = \left\{(-1)^{m\pm n}\dfrac{1}{n}\right\}$.

Example 1.5.4. Consider the partial difference equation

$$\nabla_m \nabla_n y(m,n) = \frac{2(m+k_1)}{m} \, y(m+k_1, n+\ell_1) +$$

$$+ \frac{2(m+k_2)}{m-1} \, y(m+k_2, n+\ell_2), \, m \geq 2, \, n \geq 1 \qquad (1.5.4)$$

where $(k_1 + \ell_1)$ and $(k_2 + \ell_2)$ are even.

By letting $f_1(u) = f_2(u) = u$ and

$$q_1(m,n) = q_1'(m,n) = \frac{2(m+k_1)}{m}, \quad q_2(m,n) = q_2'(m,n) = \frac{2(m+k_2)}{m-1},$$

$$p_i(m,n) = p_i'(m,n) = 0, \, i = 1,2,$$

we check that the hypotheses of Theorems 1.3.2 and 1.3.3 are satisfied. Therefore, equation (1.5.4) has no unbounded nonoscillatory solution. In fact, (1.5.4) has oscillatory solutions given by $\{y(m,n)\} = \left\{ (-1)^{m \pm n} \dfrac{1}{m} \right\}$.

Example 1.5.5. Consider the partial difference equation

$$\nabla_m \nabla_n y(m,n) = \frac{(2n-1)(m+k_1)(n+\ell_1)}{mn(n-1)} \, y(m+k_1, n+\ell_1)$$

$$+ \frac{(2n-1)(m+k_2)(n+\ell_2)}{n(n-1)(m-1)} \, y(m+k_2, n+\ell_2), \, m,n \geq 2 \qquad (1.5.5)$$

where $(k_1 + \ell_1)$ and $(k_2 + \ell_2)$ are even.

We take $f_1(u) = f_2(u) = u$ and

$$q_1(m,n) = q_1'(m,n) = \frac{(2n-1)(m+k_1)(n+\ell_1)}{mn(n-1)},$$

$$q_2(m,n) = q_2'(m,n) = \frac{(2n-1)(m+k_2)(n+\ell_2)}{n(n-1)(m-1)},$$

$$p_i(m,n) = p_i'(m,n) = 0, \, i = 1,2.$$

It follows that $q_i - p_i' = 2, \, i = 1,2$.
Clearly, (1.3.2) is fulfilled.
Further, condition (1.3.9) holds as

$$\limsup_{m,n \to \infty} \sum_{s=1}^{2} \gamma_s \sum_{i=m-k'}^{m} \sum_{j=n-\ell'}^{n} [q_s(i,j) - p_s'(i,j)] = 4(k'+1)(\ell'+1) > \ell'+2.$$

We conclude from Theorems 1.3.2 and 1.3.3, that equation (1.5.5) has no unbounded nonoscillatory solution. In fact, (1.5.5) has oscillatory solutions given by $\{y(m,n)\} = \left\{(-1)^{m \pm n} \dfrac{1}{mn}\right\}$.

Example 1.5.6. Consider the partial difference equation

$$\nabla_m \nabla_n y(m,n) = \frac{(m+k)^j(n+\ell)^j}{m(m-1)n(n-1)} [y(m+k,n+\ell)]^j, \ m,n \geq 2 \qquad (1.5.6)$$

where $j \ (\geq 3)$ is any odd integer.

Taking $f(u) = u^j$, we may choose

$$q(m,n) = q'(m,n) = \frac{(m+k)^j(n+\ell)^j}{m(m-1)n(n-1)}, \quad p(m,n) = p'(m,n) = 0.$$

Thus, $\lim_{m,n\to\infty}[q(m,n) - p'(m,n)] = \infty$. It can easily be checked that all the conditions of Corollaries 1.2.2 - 1.2.6, Theorems 1.3.2 and 1.3.3 are fulfilled. Hence, (1.5.6) has no unbounded nonoscillatory solution. In fact, (1.5.6) has an eventually positive and decreasing (i.e., bounded nonoscillatory) solution given by $\{y(m,n)\} = \left\{\dfrac{1}{mn}\right\}$.

Example 1.5.7. Consider the partial difference equation

$$\nabla_m \nabla_n y(m,n) = \frac{1}{(m+k)(n+\ell)} \, y(m+k,n+\ell), \ m,n \geq 1. \qquad (1.5.7)$$

Let $f(u) = u$ and

$$q(m,n) = q'(m,n) = \frac{1}{(m+k)(n+\ell)}, \quad p(m,n) = p'(m,n) = 0.$$

Since $\lim_{m,n\to\infty}[q(m,n) - p'(m,n)] = 0$, the hypothesis (A3) is violated. In fact, equation (1.5.7) has a family of eventually positive and increasing solutions given by $\{y(m,n)\} = \{(m+a)(n+b)\}$ where $a,b \geq 0$.

Example 1.5.8. Consider the partial difference equation

$$\Delta_m \Delta_n y(m,n) + \frac{2\beta(2n+1)}{n(n+1)} \, y(m-\alpha, n-\beta)$$

$$= \frac{2(2n+1)}{n+1} \, y(m-\alpha, n-\beta), \quad m \geq \alpha, \ n \geq \beta+1 \qquad (1.5.8)$$

where α and β are nonnegative odd/even integers.

Here, $\tau = 1$, $g_1(m) = m - \alpha$ and $h_1(n) = n - \beta$. Choosing $f_1(u) = u$, we have $\gamma_1 = 1$. Further, since

$$\frac{P_1(m, n, y(g_1(m), h_1(n)))}{f_1(y(g_1(m), h_1(n)))} = \frac{2\beta(2n + 1)}{n(n + 1)},$$

$$\frac{Q_1(m, n, y(g_1(m), h_1(n)))}{f_1(y(g_1(m), h_1(n)))} = \frac{2(2n + 1)}{n + 1},$$

we may take

$$p_1(m, n) = p_1'(m, n) = \frac{2\beta(2n + 1)}{n(n + 1)}, \quad q_1(m, n) = q_1'(m, n) = \frac{2(2n + 1)}{n + 1}.$$

Thus, (C1)–(C3) are fulfilled. It is also noted that

$$\mu_1(m, n) = q_1(m, n) - p_1'(m, n) = 2\left(2 + \frac{1}{n}\right)\left(1 - \frac{\beta + 1}{n + 1}\right)$$

$$\geq 4\left(1 - \frac{\beta + 1}{\beta + 2}\right) = \frac{4}{\beta + 2} \equiv \rho_1.$$

Case 1: Theorem 1.4.12

Condition (1.4.19) is the same as

$$\frac{\alpha + \beta + 1}{\alpha + \beta}\left[2^r(\alpha + \beta)\frac{4}{\beta + 2}\right]^{1/(\alpha + \beta + 1)} > 2. \tag{1.5.9}$$

Hence, (1.5.8) has neither eventually positive and nonincreasing nor eventually negative and nondecreasing solutions if α and β satisfy (1.5.9). Examples of such α and β include $(\alpha, \beta) = (3, 1)$, $(2, 2)$. In fact, for any odd/even α and β, we find that (1.5.8) has oscillatory solutions given by $\{y(m, n)\} = \left\{(-1)^{m \pm n}\frac{1}{n}\right\}$.

Case 2: Theorem 1.4.13

When $\alpha = \beta = 0$, we find that $\mu_1(m, n) \geq \rho_1 = 2$ and so (1.4.23) is fulfilled. Thus, the conclusion of Theorem 1.4.12(iii) follows.

Case 3: Theorem 1.4.14

We have

$$\limsup_{M,N \to \infty} \sum_{m=M}^{M+\alpha} \sum_{n=N}^{N+\beta} \gamma_1 \mu_1(m, n) = 4(\alpha + 1)(\beta + 1) > 1$$

and so (1.4.25) is satisfied. Hence, the conclusion of Theorem 1.4.12(iii) holds for all odd/even α and β.

Case 4: Theorem 1.4.15

Since

$$\sum_{m=M-\alpha}^{M-1} \sum_{n=N-\beta}^{N-1} \gamma_1 \mu_1(m,n) = \alpha \sum_{n=N-\beta}^{N-1} 2\left(2+\frac{1}{n}\right)\left(1-\frac{\beta+1}{n+1}\right)$$

$$\geq \alpha \sum_{n=N-\beta}^{N-1} 2\left(2+\frac{1}{N-1}\right)\left(1-\frac{\beta+1}{N-\beta+1}\right)$$

$$= 2\alpha\beta\left(2+\frac{1}{N-1}\right)\left(1-\frac{\beta+1}{N-\beta+1}\right) \geq 2\alpha\beta$$

for all large N, we see that condition (1.4.10) holds for all large M and N when $b=\frac{1}{2}$. With $c=4$, it is clear that condition (1.4.27) reduces to

$$4(\alpha+1)(\beta+1) > 15$$

which is obviously true for any $\alpha, \beta > 0$. Therefore, we conclude as in Case 3.

Case 5: Theorem 1.4.16

From Case 4, it is obvious that (1.4.10) holds for all large M and N when $b=2$. Thus, condition (1.4.30) is fulfilled and we conclude as in Case 3.

Case 6: Theorem 1.4.17

Let $\beta=0$, $\alpha>0$. Then,

$$\sum_{m=M-\alpha}^{M-1} \gamma_1 \mu_1(m,N) = \sum_{m=M-\alpha}^{M-1} \frac{2(2N+1)}{N+1} \geq 2\alpha$$

and so (1.4.13) holds when $b=\frac{1}{2}$. With $c=4$, it is clear that

$$\limsup_{M,N\to\infty} \sum_{m=M-\alpha}^{M} \gamma_1 \mu_1(m,N) = 4(\alpha+1) > 1-\frac{1}{c}.$$

Therefore, (1.4.31) is satisfied and the conclusion of Theorem 1.4.12(iii) holds for all even $\alpha > 0$.

Case 7: Theorem 1.4.18

Let $\alpha=0$, $\beta>0$. Since

$$\sum_{n=N-\beta}^{N-1} \gamma_1 \mu_1(M,n) = \sum_{n=N-\beta}^{N-1} \frac{2(2n+1)}{n+1} \geq 2\beta,$$

we find that (1.4.15) holds when $b = \frac{1}{2}$. With $c = 4$, it is obvious that (1.4.34) is fulfilled. Thus, the conclusion of Theorem 1.4.12(iii) holds for all even $\beta > 0$.

Case 8: Theorem 1.4.19

Let $\beta = 0$, $\alpha > 0$. From Case 6, it is clear that (1.4.13) holds for $b = 2\alpha$. It follows that (1.4.35) is satisfied as the left side is $4(\alpha + 1)(> 0)$ and the right side is negative. Hence, we conclude as in Case 6.

Case 9: Theorem 1.4.20

Let $\alpha = 0$, $\beta > 0$. From Case 7, it is clear that (1.4.15) holds for $b = 2\beta$. Therefore, condition (1.4.36) is fulfilled and we conclude as in Case 7.

Case 10: Theorem 1.4.21

The condition (1.4.37) reduces to

$$4 > w. \tag{1.5.10}$$

Thus, the conclusion of Theorem 1.4.12(iii) holds for those α and β that fulfill (1.5.10), e.g., $(\alpha, \beta) = (2, 4)$, $(4, 2)$, $(1, 5)$, $(5, 1)$.

Case 11: Theorem 1.4.22

Let $\beta = 0$, $\alpha > 0$. The left side of (1.4.47) is 4 which is of course more than the right side. Hence, we conclude as in Case 6.

Case 12: Theorem 1.4.23

Let $\alpha = 0$, $\beta > 0$. Again, the left side of (1.4.54) is 4 which is of course more than the right side. Thus, we conclude as in Case 7.

Case 13: Theorem 1.4.24

Here, we have $v_1 = 3$ and condition (1.4.56) is the same as

$$\limsup_{m,n \to \infty} \gamma_1 \mu_1(m, n) + 2v_1 = 4 + 2(3) = 10 > 4 - 2\tau = 4 - 2 = 2.$$

Therefore, we conclude as in Case 3.

Case 14: Theorems 1.4.26–1.4.28

We have $\mu_1 = 4$. Subsequently, condition (1.4.62) reduces to

$$\frac{(r_1 + 1)^{r_1+1}}{r_1^{r_1}} > \frac{1}{4} \tag{1.5.11}$$

which is obviously true. It can be checked that conditions (1.4.67) and (1.4.68) also reduce to (1.5.11). Thus, we conclude as in Case 3.

Example 1.5.9. Consider the partial difference equation

$$\Delta_m \Delta_n y(m,n) = \frac{2(2m+1)(2n+1)(m-\alpha_1)(n-\beta_1)}{5mn(m+1)(n+1)} \, y(m-\alpha_1, n-\beta_1)$$

$$+ \frac{3(2m+1)(2n+1)(m-\alpha_2)(n-\beta_2)}{5mn(m+1)(n+1)} \, y(m-\alpha_2, n-\beta_2),$$

$$m \geq \alpha_1 + \alpha_2 + 1, \; n \geq \beta_1 + \beta_2 + 1 \tag{1.5.12}$$

where $\alpha_i, \beta_i, \; i = 1, 2$ are nonnegative odd/even integers.

In this example, $\tau = 2$, $g_i(m) = m - \alpha_i$, $h_i(n) = n - \beta_i$, $i = 1, 2$. Taking $f_1(u) = f_2(u) = u$, we have $\gamma_1 = \gamma_2 = 1$. Further, we let

$$q_1(m,n) = q_1'(m,n) = \frac{2(2m+1)(2n+1)(m-\alpha_1)(n-\beta_1)}{5mn(m+1)(n+1)},$$

$$q_2(m,n) = q_2'(m,n) = \frac{3(2m+1)(2n+1)(m-\alpha_2)(n-\beta_2)}{5mn(m+1)(n+1)},$$

$$p_1(m,n) = p_1'(m,n) = p_2(m,n) = p_2'(m,n) = 0$$

so that (C1)–(C3) are fulfilled. Subsequently, it is noted that

$$\mu_1(m,n) = \frac{2}{5}\left(2 - \frac{1}{m+1}\right)\left(2 - \frac{1}{n+1}\right)\left(1 - \frac{\alpha_1}{m}\right)\left(1 - \frac{\beta_1}{n}\right)$$

$$\equiv z(\alpha_1, \beta_1, m, n) \geq z(\alpha_1, \beta_1, \alpha_1 + \alpha_2 + 1, \beta_1 + \beta_2 + 1) \equiv \rho_1$$

and

$$\mu_2(m,n) = \frac{3}{2} \, z(\alpha_2, \beta_2, m, n) \geq \frac{3}{2} \, z(\alpha_2, \beta_2, \alpha_1 + \alpha_2 + 1, \beta_1 + \beta_2 + 1) \equiv \rho_2.$$

Case 1: Theorem 1.4.12

Equation (1.5.12) has no eventually positive (negative) and nonincreasing (nondecreasing) solution if (1.4.19) is satisfied. As an example, (1.4.19) is fulfilled when $\alpha_1 = \beta_1 = 3$, $\alpha_2 = 4$ and $\beta_2 = 2$. In fact, it is noted that for any odd/even $\alpha_i, \beta_i, \; i = 1, 2$, the equation (1.5.12) has oscillatory solutions given by

$$\{y(m,n)\} = \left\{ (-1)^{m \pm n} \frac{1}{mn} \right\}.$$

Case 2: Theorem 1.4.21

We find that condition (1.4.37) is the same as (1.5.10), which is satisfied by, for example $\alpha_1 = 2$, $\beta_1 = 4$, $\alpha_2 = 3$ and $\beta_2 = 5$.

Case 3: Theorems 1.4.13–1.4.20, 1.4.22–1.4.24, 1.4.26–1.4.28

It can be checked that the conditions of the above theorems are satisfied for any α and β. Thus, the conclusion of Theorem 1.4.12(iii) holds for any odd/even α_i, β_i, $i = 1, 2$.

Example 1.5.10. Consider the partial difference equation

$$\Delta_m \Delta_n y(m, n) = (2n + 1)[(m - \alpha)^i][(n - \beta)^j]^2 \left\{ y([(m - \alpha)^i], [(n - \beta)^j]) \right\}^{-1},$$

$$m \geq \alpha + 1, \ n \geq \beta + 1 \tag{1.5.13}$$

where $[\cdot]$ is the greatest integer function, $0 < i, j \leq 1$ and α, β are nonnegative integers.

Here, $\tau = 1$, $g_1(m) = [(m - \alpha)^i] \leq m - \alpha$ and $h_1(n) = [(n - \beta)^j] \leq n - \beta$. Choosing $f_1(u) = 1/u$, we may take

$$p_1(m, n) = p_1'(m, n) = 0, \quad q_1(m, n) = q_1'(m, n) = (2n + 1)[(m - \alpha)^i][(n - \beta)^j]^2.$$

Thus, $\mu_1(m, n) \to \infty$ as $m, n \to \infty$. It can be easily checked that all the conditions of Theorems 1.4.12–1.4.28 are satisfied. Hence, (1.5.13) has neither eventually positive and nonincreasing nor eventually negative and nondecreasing solutions. Indeed, we note that (1.5.13) has an eventually positive and increasing solution given by $\{y(m, n)\} = \{mn^2\}$.

Example 1.5.11. Consider the partial difference equation

$$\Delta_m \Delta_n y(m, n) = \frac{1}{4}(m - \alpha_1)[(n - \beta_1)^{1/2}] \left\{ y(m - \alpha_1, [(n - \beta_1)^{1/2}]) \right\}^{-1}$$

$$+ \frac{3}{4}[(m - \alpha_2)^{1/3}](n - \beta_2) \left\{ y([(m - \alpha_2)^{1/3}], n - \beta_2) \right\}^{-1},$$

$$m \geq \alpha_1 + \alpha_2 + 1, \ n \geq \beta_1 + \beta_2 + 1 \tag{1.5.14}$$

where $[\cdot]$ is the greatest integer function and α_i, β_i, $i = 1, 2$ are nonnegative integers.

Clearly, $g_1(m) = m - \alpha_1$, $h_1(n) = [(n - \beta_1)^{1/2}] \leq n - \beta_1$, $g_2(m) = [(m - \alpha_2)^{1/3}] \leq m - \alpha_2$ and $h_2(n) = n - \beta_2$. We shall take $f_1(u) = f_2(u) = 1/u$ and

$$p_1(m, n) = p_1'(m, n) = 0, \quad q_1(m, n) = q_1'(m, n) = \frac{1}{4}(m - \alpha_1)[(n - \beta_1)^{1/2}],$$

$$p_2(m, n) = p_2'(m, n) = 0, \quad q_2(m, n) = q_2'(m, n) = \frac{3}{4}[(m - \alpha_2)^{1/3}](n - \beta_2).$$

Subsequently, $\mu_i(m,n) \to \infty$, $i = 1,2$ as $m,n \to \infty$. Again it can easily be checked that all the conditions of Theorems 1.4.12–1.4.24, 1.4.26–1.4.28 are fulfilled. Hence, we conclude as in Example 1.5.10. Indeed, equation (1.5.14) has an eventually positive and increasing solution given by $\{y(m,n)\} = \{mn\}$.

Example 1.5.12. Consider the partial difference equation

$$\Delta_m \Delta_n y(m,n) = \frac{(m - \alpha_1)(n - \beta_1)}{3mn(m+1)(n+1)} \, y(m - \alpha_1, n - \beta_1)$$

$$+ \frac{2(m - \alpha_2)(n - \beta_2)}{3mn(m+1)(n+1)} \, y(m - \alpha_2, n - \beta_2),$$

$$m \geq \alpha_1 + \alpha_2 + 1, \ n \geq \beta_1 + \beta_2 + 1 \tag{1.5.15}$$

where α_i, β_i, $i = 1,2$ are nonnegative integers.

Here, $g_i(m) = m - \alpha_i$, $h_i(n) = n - \beta_i$, $i = 1,2$. Further, we let $f_1(u) = f_2(u) = u$ and

$$p_1(m,n) = p_1'(m,n) = 0, \quad q_1(m,n) = q_1'(m,n) = \frac{(m - \alpha_1)(n - \beta_1)}{3mn(m+1)(n+1)},$$

$$p_2(m,n) = p_2'(m,n) = 0, \quad q_2(m,n) = q_2'(m,n) = \frac{2(m - \alpha_2)(n - \beta_2)}{3mn(m+1)(n+1)}.$$

Subsequently, $\mu_i(m,n) \to 0$, $i = 1,2$ as $m,n \to \infty$. It can be easily checked that all the conditions of Theorems 1.4.12–1.4.24, 1.4.26–1.4.28 are violated. Hence, we cannot conclude that (1.5.15) has no eventually positive (negative) and non-increasing (nondecreasing) solution. In fact, (1.5.15) does have an eventually positive and nonincreasing solution given by $\{y(m,n)\} = \left\{ \dfrac{1}{mn} \right\}$.

References

[1] R. P. Agarwal. *Difference equations and inequalities*. Marcel Dekker, 1992.

[2] R. P. Agarwal. Difference equations and inequalities: A survey. *Proceedings of the First World Congress on Nonlinear Analysts 1992, ed. V. Lakshmikantham, Walter de Gruyter and Co.*, pages 1091–1108.

[3] R. P. Agarwal, M. M. S. Manuel, and E. Thandapani. Oscillatory and nonoscillatory behavior of second order neutral delay difference equations. *Math. Comput. Modelling*, 24:5–11, 1996.

[4] R. P. Agarwal, S. Pandian, and E. Thandapani. Oscillatory property for second order nonlinear difference equations via lyapunov second method. *Advances in Nonlinear Dynamics*, Stability Control Theory Methods Appl. 5, Gordon and Breach, Amsterdam:11–21, 1996.

[5] R. P. Agarwal, E. Thandapani, and P. J. Y. Wong. Oscillations of higher order neutral difference equations. *Appl. Math. Lett.*, 10:71–78, 1997.

[6] S. S. Cheng and W. T. Patula. An existence theorem for a nonlinear difference equation. *Nonlinear Analysis*, 20:193–203, 1993.

[7] I. Györi and G. Ladas. *Oscillation Theory of Delay Differential Equations with Applications*. Clarendon Press, Oxford, 1991.

[8] W. G. Kelley and A.C. Peterson. *Difference Equations: An Introduction with Applications*. Academic Press, New York, 1991.

[9] V. L. Kocic and G. Ladas. *Global Behavior of Nonlinear Difference Equations of Higher Order with Applications*. Kluwer, Dordrecht, 1993.

[10] V. Lakshmikantham and D. Trigiante. *Difference Equations with Applications to Numerical Analysis*. Academic Press, New York, 1988.

[11] J. J. Li and C. C. Yeh. Existence of positive nondecreasing solutions of nonlinear difference equations. *Nonlinear Analysis*, 22:1271–1284, 1994.

[12] J. Popenda. The oscillation of solutions of difference equations. *Comput. Math. Appl.*, 28:271–279, 1994.

[13] A. N. Sharkovsky, Y. L. Maistrenko, and E. Y. Romanenko. *Difference equations and their applications*. Kluwer Academic Publishers, 1993.

[14] E. Thandapani. Oscillation theorems for perturbed nonlinear second order difference equations. *Comput. Math. Appl.*, 28:309–316, 1994.

[15] P. J. Y. Wong and R. P. Agarwal. Oscillation theorems and existence of positive monotone solutions for second order nonlinear difference equations. *Math. Comput. Modelling*, 21:63–84, 1995.

[16] P. J. Y. Wong and R. P. Agarwal. Oscillation theorems for certain second order nonlinear difference equations. *J. Math. Anal. Appl*, 204:813–829, 1996.

[17] P. J. Y. Wong and R. P. Agarwal. Oscillation and monotone solutions of second order quasilinear difference equations. *Funkcialaj Ekvacioj*, 39:491–517, 1996.

[18] P. J. Y. Wong and R. P. Agarwal. On the oscillation of an mth order perturbed nonlinear difference equation. *Archivum Mathematicum*, 32:13–27, 1996.

[19] P. J. Y. Wong and R. P. Agarwal. Summation averages and the oscillation of second order nonlinear difference equations. *Math. Comput. Modelling*, 24:21–35, 1996.

[20] P. J. Y. Wong and R. P. Agarwal. Comparison theorems for the oscillation of higher order difference equations with deviating arguments. *Math. Comput. Modelling*, 24:39–48, 1996.

[21] P. J. Y. Wong and R. P. Agarwal. Oscillation and nonoscillation of half-linear difference equations generated by deviating arguments. *Comput. Math. Appl.*, 36:11–26, 1998.

[22] S. S. Cheng, S. L. Xie, and B. G. Zhang. Qualitative theory of partial difference equations (ii): Oscillation criteria for direct control systems in several variables. *Tamkang J. Math.*, 26:65–79, 1995.

[23] S. S. Cheng, S. L. Xie, and B. G. Zhang. Qualitative theory of partial difference equations (iii): Forced oscillations of parabolic type partial difference equations. *Tamkang J. Math.*, 26:177–192, 1995.

[24] S. S. Cheng, S. L. Xie, and B. G. Zhang. Qualitative theory of partial difference equations (iv): Forced oscillations of hyperbolic type nonlinear partial difference equations. *Tamkang J. Math.*, 26:337–360, 1995.

[25] S. S. Cheng, S. L. Xie, and B. G. Zhang. Qualitative theory of partial difference equations (v): Sturmian theorems for a class of partial difference equations. *Tamkang J. Math.*, 27:89–97, 1996.

[26] S. S. Cheng and B. G. Zhang. Qualitative theory of partial difference equations (i): Oscillation of nonlinear partial difference equations. *Tamkang J. Math.*, 25:279–288, 1994.

[27] S. C. Fu and L. Y. Tsai. Oscillation in nonlinear difference equations. *Comput. Math. Appl.*, 36:193–201, 1998.

[28] X. P. Li. Partial difference equations used in the study of molecular orbits. *Acta Chimica Sinica*, 40:688–698, 1982.

[29] S. H. Saker and P. J. Y. Wong. Nonexistence of unbounded nonoscillatory solutions of nonlinear perturbed partial difference equations. *J. Concrete and Applicable Mathematics*, 1:87–99, 2003.

[30] P. J. Y. Wong and R. P. Agarwal. Oscillation criteria for nonlinear partial difference equations with delays. *Comput. Math. Appl.*, 32:57–86, 1996.

[31] P. J. Y. Wong and R. P. Agarwal. Nonexistence of unbounded nonoscillatory solutions of partial difference equations. *J. Math. Anal. Appl.*, 214:503–523, 1997.

[32] P. J. Y. Wong. Eventually positive and monotonely decreasing solutions of partial difference equations. *Comput. Math. Appl.*, 35:35–58, 1998.

[33] P. J. Y. Wong and R. P. Agarwal. On the oscillation of partial difference equations generated by deviating arguments. *Acta Mathematica Hungarica*, 79:1–29, 1998.

[34] P. J. Y. Wong and R. P. Agarwal. Asymptotic behaviour of solutions of higher order difference and partial difference equations with distributed deviating arguments. *Appl. Math. Comput*, 97:139–164, 1998.

[35] P. J. Y. Wong. Asymptotic behaviour of solutions of partial difference inequalities. *Proceedings of the Conference on Differential Equations and their Applications (EQUADIFF 9)*, 1998.

[36] B. G. Zhang and S. T. Liu. Oscillation of partial difference equations. *PanAmerican Math. J.*, 5:61–70, 1995.

[37] B. G. Zhang and S. T. Liu. Oscillation of partial difference equations with variable coefficients. *Comput. Math. Appl.*, 36:235–242, 1998.

[38] B. G. Zhang, S. T. Liu, and S. S. Cheng. Oscillation of a class of delay partial difference equations. *J. Difference Equ. Appl.*, 1:215–226, 1995.

[39] J. C. Strikwerda. *Finite Difference Schemes and Partial Differential Equations*. Wadsworth and Brooks, 1989.

CHAPTER 2

Functional-analysis and partial difference equations

Eugenia N. Petropoulou[1] and Panayiotis D. Siafarikas[2,1]

[1]**Department of Engineering Sciences, Division of Applied Mathematics and Mechanics, University of Patras, 26500 Patras, Greece.**
e-mail: jenpetro@des.upatras.gr
and
[2]**Department of Mathematics, Division of Applied Analysis, University of Patras, 26500 Patras, Greece.**
e-mail: panos@math.upatras.gr

Abstract: In this review paper, the aim is to present a functional–analytic method developed relevantly recently by the authors for the study of partial difference equations in the spaces ℓ_1 and ℓ_2. The method is demonstrated using two illustrative examples. At the same time an effort is made in order to present several other methods used by other researchers, which make use of functional analysis and operator theory.

Mathematics Subject Classification. 39-02, 39A05, 39A11, 47N99

Key words and phrases: partial difference equations, functional analysis, boundedness, asymptotic stability, region of attraction, oscillation

2.1 Introduction.

Partial difference equations (PΔEs) arise naturally in several mathematical or realistic problems and thus their study is necessary for understanding and/or solving these problems. To begin with, PΔEs were initially considered only as

---
[1]Deceased on 26-06-2010

the discrete analogue of partial differential equations (PDEs) and as a consequence, several methods for their study have been developed which resemble standard methods of PDEs, such as the separation of variables method, or the z−transform method (which takes after, in philosophy, the Laplace or Fourier transform method for PDEs). (See for example [1–5]).

Partial difference equations appear also naturally, when discretizing PDEs using various numerical schemes (standard or non–standard) in order to solve them numerically (see for example [6–8]). This fact, together with the tremendous developments of computers, have given a new impulse in the need for research in the quite recently newly developed area of PΔEs.

Of course, nowadays, PΔEs are considered as a mathematical topic of each own which has some few similarities with PDEs, but also many differences with them. These differences are even stronger when it comes to non–linear equations. At this point one could quote the words of P. Lax in [9]:

"...This brief article is about the numerical solution of partial differential equations. A very general (although by no means the only) method for solving these is to convert them into difference equations through the replacement of derivatives by difference quotients of some sort...My aim is to convince a skeptical reader who may regard using finite differences as the last resort of a scoundrel that the theory of difference equations is a rather sophisticated affair, more sophisticated than the corresponding theory of partial differential equations. My argument will be based on two contentions:

1) In order to prove that solutions of a sequence of difference equations converge one needs estimates for difference equations operators which are analogous to the estimates needed in the existence and uniqueness theory for solutions of differential equations.

2) Estimates for difference operators are much harder to derive than the corresponding estimates for differential operators"

and the words of A. Sharkovsky, Yu. Maistrenko and E. Romanenko in the introduction of [10]:

"...Many statements concerning the theory of linear differential equations are also valid for the corresponding difference equations...The distinctive features of difference equations, as compared with differential equations, manifest themselves most impressively when the difference equations are nonlinear. In this case, their effective investigation involves the development of special–purpose approaches and techniques."

From the point of view of applications which involve PΔEs, one could mention many problems such as mathematical, physical, economical, chemical, medical, as well as problems in dynamics of populations, game theory, queuing theory, etc. Returning now to the research mathematical problems studied for PΔEs and the mathematical methods used, it would be useful to summarize some of them. First of all the majority of research papers in PΔEs deal with

(P1) the existence and uniqueness of solutions in various spaces,

(P2) the existence of solutions in explicit form,

(P3) the existence of oscillatory solutions,

(P4) the existence of positive (negative) solutions,

(P5) the existence of periodic solutions,

(P6) the asymptotic behavior of their solutions,

(P7) the growth or boundedness of their solutions,

(P8) eigenvalue problems,

(P9) integrability problems,

(P10) boundary value problems.

On the other hand, the methods used vary from methods involving techniques of multi–dimensional calculus, to methods using functional analysis, fixed point theorems and operator theory. It is the aim of the present paper to present a functional–analytic method developed by the authors for the study of PΔEs of two variables in the spaces

$$\ell_1 = \left\{ f(i,j) : \mathbb{N} \times \mathbb{N} \to \mathbb{C} \text{ where } \sum_{i=1}^{\infty} \sum_{j=1}^{\infty} |f(i,j)| < \infty \right\} \qquad (2.1.1)$$

and

$$\ell_2 = \left\{ f(i,j) : \mathbb{N} \times \mathbb{N} \to \mathbb{C} \text{ where } \sum_{i=1}^{\infty} \sum_{j=1}^{\infty} |f(i,j)|^2 < \infty \right\}, \qquad (2.1.2)$$

as well as to briefly present other similar in philosophy techniques (see §2). (A generalization for PΔEs of n variables is also available, see [11].) In §2 also, several advantages of using functional analysis, fixed point theorems and operator theory in the study of PΔEs will be given. In §4, the method developed by the authors will be presented and illustrated in §5 and §6 by use of two examples. Moreover in §3, several motivations will be given for studying PΔEs in the spaces ℓ_1 and ℓ_2 defined in (2.1.1) and (2.1.2), respectively.

2.2 Why and how to use functional analysis?

In this section the advantages of using functional analysis, as well as several methods based on functional analysis, in the study of PΔEs will be presented. First of all it should be clarified, that by referring to methods using functional analysis (in this paper) it is meant, methods using all kind of techniques and notations widely used in functional analysis and operator theory, as well as methods which make use of fixed point theorems.

One could admit that there is a "debate" on why using functional–analytic techniques instead of, for example, techniques of calculus. There are of course several advantages and disadvantages of both these categories of methods. For example when using functional–analytic methods, the proofs are (generally speaking) more clear and general, in the sense that the same ideas or techniques could be used for the study of difference equations (or other similar problems even for other types of equations). Also it could be easier to generalize results regarding PΔEs in $\mathbb{R}$ to $\mathbb{C}$. On the other hand, it is possible that a result concerning a PΔE in $\mathbb{C}$ is much weaker than the corresponding result in $\mathbb{R}$ when proving it by using functional–analytic techniques, but when using "classical" techniques (for example making use of certain inequalities) the same result turns out to be much stronger. As always in mathematics (and life) no technique is perfect!

However, it is quite acceptable that functional analysis provides a common language and notation for denoting certain types of the same mathematical arguments appearing in various scientific topics. Some times, the specific scientific topics are so full with details that one may not recognize at once that the mathematical argument is the same. Functional analysis "eliminates" any non essential details and emerges the principal arguments. At this point the works of B. Noble [12] and L. Rall [13] should be mentioned, where one can find why functional analysis is important in research. Although these papers were written 35 years ago, they remain very modern and up to date. Actually, the whole volume: "Applied Functional Analysis in Teaching and Research", IMA Bulletin, Vol. 10, No. 2, April 1974, contains many papers showing how functional analysis can successfully be applied in order to deal with such diverse topics as hydrodynamics, numerical solutions of PDEs, optimization and thermodynamics.

In this paper, it is the intention of the authors to show how functional analysis, operator theory and fixed point theorems can be very useful in the study of PΔEs. To begin with, an outline of the functional–analytic method developed by the authors for the study of PΔEs, which will be presented in details in §4, is given. This method provides conditions for the existence of solutions of PΔEs in the spaces ℓ_1 and ℓ_2 defined in (2.1.1) and (2.1.2), respectively. In this way, problems of (P1) type are dealt. Also, due to the nature of the spaces ℓ_1 and ℓ_2 information is given regarding the asymptotic behavior of the solutions of PΔEs (problems of (P6) type). Moreover, this method has a constructive character, which enables finding regions where these solutions hold, as well as bounds of them (problems of (P7) type). Finally, the method was recently applied to an eigenvalue problem for linear PΔEs, which apart from information regarding the eigenvalues of the problem, information concerning the behavior of its eigenfunctions (positivity, oscillation) were deduced. This is a quite promising fact, that this method can also be applied for the study of problems of (P3) or (P8) type. (For more details about the history and evolution of the method, see §4.)

The main idea of this method is to transform the PΔE under consideration into an equivalent operator equation in an abstract Hilbert (H) or Banach space (H_1), by the use of an isomorphism between ℓ_2 and H or ℓ_1 and H_1. In this way, the study of the initial PΔE is reduced to the study of an equivalent operator equation in which the initial or boundary conditions accompanying the PΔE are incorporated.

Of course, there are other researchers who also use techniques or tools of functional analysis and operator theory in their studies on PΔEs. Some relevant papers are mentioned indicatively, but the reference list of this paper is by no means exhaustive. In the work of B. Zhang and Y.Zhou in [5] (see also the references therein), several fixed point theorems such as the Knaster–Tarksi, Schauder and Banach fixed point theorems, have been used in order to prove results concerning the boundedness, positivity and oscillatory or non–oscillatory behavior of the solutions of several classes of linear and non–linear PΔEs. In order to apply these fixed point theorems, the authors work on specific spaces of double sequences with specific norms that are adequate for their type of results and, construct certain mappings which satisfy the criteria for the application of the above mentioned fixed point theorems. Also in [14], the authors apply the Banach contraction principle in order to prove the existence of bounded and/or positive solutions of a non–linear partial difference equation.

Finally in [15–19], the authors define an appropriate Banach space and a cone and then apply several appropriate fixed point theorems, such as the Krasnosel'skii fixed point theorem or others that can be found in [20–22]. In this way, they obtain results concerning the existence of positive radial or positive symmetric radial solutions of certain boundary value problems for partial difference equations.

2.3   Why to study partial difference equations in ℓ_1, ℓ_2

As already mentioned, the functional–analytic method that will be presented in details in §4, deals with the study of PΔEs in the spaces ℓ_1 and ℓ_2, defined by (2.1.1) and (2.1.2), respectively. The motivation, for seeking solutions of PΔEs in ℓ_1 and ℓ_2 arises from various mathematical and physical problems described in the following:

1) **General motivation arising from the definitions of ℓ_1, ℓ_2.**

It is obvious from (2.1.1) and (2.1.2), that the spaces ℓ_1 and ℓ_2 can be considered as the discrete analogues of the spaces

$$L_1 = \left\{ f : \mathbb{C} \times \mathbb{C} \to \mathbb{C} \text{ where } \int\int_D |f(x,y)| dx dy < \infty \right\}$$

and

$$L_2 = \left\{ f : \mathbb{C} \times \mathbb{C} \to \mathbb{C} \text{ where } \int\int_D |f(x,y)|^2 dx dy < \infty \right\},$$

where D is a region of $\mathbb{C} \times \mathbb{C}$, a fact which is quite interesting on its own. Also, it is obvious from (2.1.1) and (2.1.2) that $\lim_{i,j\to\infty} f(i,j) = 0$ and as a consequence solutions of PΔEs in ℓ_1 and ℓ_2 contain information on their asymptotic behavior.

2) **Integral equations.**

In the theory of integral equations, there is developed by Hilbert a theory of functions of an infinite number of variables x_1, x_2,... of quadratic forms $\sum_{i=1}^{\infty}\sum_{j=1}^{\infty} f(i,j) x_i x_j$. This theory has been proved to be useful for functions whose squares are summable, which can be defined by means of Fourier's generalized coefficients $f(i,j)$, of suitable expansion in series of orthonormal functions. For Hilbert's theory, the consideration of functions for which $f(i,j) \in \ell_2$ is of fundamental importance [23, p. 176].

3) **Partial differential equations.**

a) In [24], a connection is made between a class of linear PDEs and a class of linear PΔEs of two variables. For this class of linear PΔEs, it is assumed that their solutions are of the form $\sum_{i=0}^{\infty}\sum_{j=1}^{\infty} |f(i,j)| \rho^{i+j} < \infty$, for a positive number ρ. Obviously, if $\rho = 1$ these solutions belong to ℓ_1.

b) Often boundary value problems of PDEs are solved using the method of separation of variables. Thus, instead of solving the initial problem, one has to solve a non-coupled system of boundary value problems of ordinary differential equations. Then by use of the superposition principle the solution of the initial value problem is written in the form of a series of a complete system of orthogonal functions, where Fourier coefficients $f(i,j)$ appear, for which it is desirable to belong to ℓ_1.

4) **Numerical solution of PDEs.**

a) In many cases, PDEs can be solved numerically by using an appropriate numerical scheme described by a PΔE. Thus, if $f(x,t)$ is the solution of a

PDE, the notation $f(i,j)$ is used for the solution of the analogue PΔE at $x = x_i$ and $t = t_j$, where $1 \leq i \leq I < +\infty$, $1 \leq j \leq J < +\infty$. Also, the notation $\epsilon(i,j)$, is used for the numerical errors of such a numerical scheme. It is generally accepted that this numerical scheme is stable if

$$\lim_{i,j \to \infty} \epsilon(i,j) = 0. \tag{2.3.1}$$

In the case of numerical schemes described by linear PΔEs, their numerical errors satisfy the same PΔE with $f(i,j)$. Thus, if it is assured that these equations (for $\epsilon(i,j)$) have a unique solution in ℓ_1 or ℓ_2, then condition (2.3.1) is automatically fulfilled.

b) In [9], Lax studied the stability of three different numerical schemes of the same linear PDE of two variables, which arise after using three different discretizations of the PDE. Among the norms used in [9], was also the ℓ_2 norm: $\|f(i,j)\|_{\ell_2} = \sum \sum |f(i,j)|^2$.

5) **Generating functions and Laurent or z$-$transforms.**

The generating function $F(z_1, z_2)$ of a double sequence $f(i,j)$ is defined by

$$F(z_1, z_2) = \sum_{i=0}^{\infty} \sum_{j=0}^{\infty} f(i,j) z_1^i z_2^j.$$

The Laurent or z$-$transform $Z(f(i,j))$ of a double sequence $f(i,j)$ is defined by

$$Z(f(i,j)) = U(w_1, w_2) = \sum_{i=0}^{\infty} \sum_{j=0}^{\infty} f(i,j) w_1^{-i} w_2^{-j}.$$

Both the generating function and the z$-$transform of a sequence, which constitute functions of two complex variables, are often used for the solution of difference equations (see for example [1]). Thus, if $f(i,j) \in \ell_1$ the convergence of the above series is guaranteed for $|z_1| < 1$, $|z_2| < 1$ and $|w_1| > 1$, $|w_2| > 1$.

6) **Perturbation methods.**

In all kind or problems described by PDEs, the solution u depends on some independent variables (for example x, y), but it may also depend on some parameters (for example π_1, π_2). When using a perturbation method in order to solve such a problem, the parameters are usually considered very small or tending to zero and one is interested in the behavior of the solution $u(x, t; \pi_1, \pi_2)$ of the problem when these parameters tend to zero and the independent variables are considered fixed. For a more exact study of the behavior of $u(x, t; \pi_1, \pi_2) = f(\pi_1, \pi_2)$, one should compare it with a function $g(\pi_1, \pi_2)$ the limiting behavior of which is known. (These comparison functions are called gauge functions). Thus one expands $f(\pi_1, \pi_2)$ in terms of a sequence of gauge functions $g_{ij}(\pi_1, \pi_2)$, i.e.:

$$f(\pi_1, \pi_2) = \sum_{i=0}^{\infty} \sum_{j=0}^{\infty} u(i, j) g_{ij}(\pi_1, \pi_2), \tag{2.3.2}$$

where $u(i, j)$ are coefficients that should be determined [25]. Thus if there exists a $M > 0$ such that

$$|g_{ij}(\pi_1, \pi_2)| \leq M$$

then the series $(2.3.2)$ converges for $u(i, j) \in \ell_1$.

7) Population dynamics.

In problems of population dynamics, the convergence conditions imposed by the definitions of ℓ_1 and ℓ_2 are quite natural and welcome.

8) Quantum mechanics.

The ℓ_2 space appears naturally in quantum mechanics. More precisely, a one–mode normalized phase state can be represented as

$$|f> = \sum_{n=0}^{\infty} f(n)|n>, \quad \sum_{n=0}^{\infty} |f(n)|^2 = 1.$$

In the same way, a two–mode normalized phase state can be represented as

$$|f> = \sum_{i=0}^{\infty} \sum_{j=0}^{\infty} f(i, j)|i, j>, \quad \sum_{n=0}^{\infty} \sum_{j=0}^{\infty} |f(i, j)|^2 = 1,$$

from which it is obvious that $f(i, j) \in \ell_2$.

Although there is a lot of motivation for studying PΔEs in ℓ_1 or ℓ_2, it should be mentioned that PΔEs involving constant or periodic non-homogeneous terms, or the problem of the existence of periodic solutions (problems of (P5) type) cannot be studied in the framework presented in this paper, since these sequences do not belong to ℓ_1 or ℓ_2.

2.4   The method in details

In this section, the functional–analytic method developed by the authors for the study of PΔEs will be presented. This method is actually a generalization of a functional–analytic method developed by E. Ifantis for ordinary difference equations, which appeared for the first time in [26] in the context of studying the spectrum of the difference equation

$$f(n+1) + f(n-1) = [E - \phi(n)]f(n).$$

However, it was not until 1987, that this method was systematically presented and used in [27], for the study of linear and non–linear ordinary difference equations. Then in [28–30] this method was further extended for other kind of non–linear ordinary difference equations and in [31], it was extended for linear systems of ordinary difference equations.

The generalization of the method for PΔEs of two variables was given in [32] for linear PΔEs and in [33] for non–linear partial PΔEs, while its generalization for PΔEs of n variables, was given in [11]. Recently, the same method was successfully applied to the eigenvalue problem of a class of linear partial difference equations in [34].

Consider the ℓ_2 space defined by (2.1.2). This space with the usual sum between two sequences and multiplication of a sequence with a complex number, equipped with the inner product

$$(f(i,j), g(i,j)) = \sum_{i=1}^{\infty} \sum_{j=1}^{\infty} f(i,j)\overline{g(i,j)}, \quad f(i,j), g(i,j) \in \ell_2. \qquad (2.4.1)$$

is a Hilbert space. The norm deduced by (2.4.1) is

$$\|f(i,j)\|_{\ell_2}^2 = \sum_{i=1}^{\infty} \sum_{j=1}^{\infty} |f(i,j)|^2.$$

Moreover, if $\{\phi_1, \phi_2, ...\}$ is an orthonormal base in the one–dimensional Hilbert space $\ell_2^{(1)}$,

$$\ell_2^{(1)} = \left\{ f(n) : \mathbb{N} \to \mathbb{C} \text{ where } \sum_{n=1}^{\infty} |f(n)|^2 < \infty \right\},$$

with $\phi_j = \{x_{jk}\}_{k=1}^{\infty}$, then the system $\{\phi_{i,j}\}_{i,j=1}^{\infty} = \{x_{ik}x_{jp}\}_{k,p=1}^{\infty}$ is an orthonormal base in ℓ_2 (see for example the not so simple exercise 70, p. 49 of [35]). In this way it is proved that ℓ_2 is a separable Hilbert space.

In a similar way, the ℓ_1 space defined by (2.1.1), with the usual sum between two sequences and multiplication of a sequence with a complex number, equipped with the norm

$$\|f(i,j)\|_{\ell_1} = \sum_{i=1}^{\infty}\sum_{j=1}^{\infty}|f(i,j)|, \quad f(i,j) \in \ell_1 \tag{2.4.2}$$

is a Banach space.

Let H denote an abstract separable Hilbert space with the orthonormal base $\{e_{i,j}\}_{i,j=1}^{\infty}$ and elements $f \in H$ which have the form $f = \sum_{i=1}^{\infty}\sum_{j=1}^{\infty}(f, e_{i,j})e_{i,j}$, with

norm $\|f\|^2 = \sum_{i=1}^{\infty}\sum_{j=1}^{\infty}|(f, e_{i,j})|^2$. Also denote by H_1 the Banach space consisting

of those elements $f \in H$ which satisfy the condition $\sum_{i=1}^{\infty}\sum_{j=1}^{\infty}|(f, e_{i,j})| < +\infty$. The

norm in H_1 is denoted by $\|f\|_1 = \sum_{i=1}^{\infty}\sum_{j=1}^{\infty}|(f, e_{i,j})|$.

Define now in H the shift operators V_1, V_2 as follows:

$$V_1 e_{i,j} = e_{i+1,j} \text{ and } V_2 e_{i,j} = e_{i,j+1}, \quad i,j = 1,2,....$$

One can easily prove that the shift operators V_1, V_2 are linear and isometric, i.e. $\|V_1 f\| = \|f\|$, $\|V_2 f\| = \|f\|$ but not unitary, i.e. their range domain is not all H. Indeed the range domain of V_1 is:

$$R(V_1) = H \ominus \{e_{1,j}, j = 1,2,3,...\}$$

and the range domain of V_2 is:

$$R(V_2) = H \ominus \{e_{i,1}, i = 1,2,3,...\}.$$

(For example, it is to see that $e_{1,1}$ does not belong to $R(V_1)$. Indeed, if $e_{1,1} \in R(V_1)$, there exists $f \in H$ such that $V_1 f = e_{1,1}$. Then

$$V_1 \sum_{i=1}^{\infty}\sum_{j=1}^{\infty}(f, e_{i,j})e_{i,j} = e_{1,1} \Rightarrow \sum_{i=1}^{\infty}\sum_{j=1}^{\infty}(f, e_{i,j})e_{i+1,j} = e_{1,1} \Rightarrow$$

$$\Rightarrow (f, e_{1,1})e_{2,1} + (f, e_{1,2})e_{2,2} + ... + (f, e_{1,k})e_{2,k} + ...+$$

$$+(f, e_{2,1})e_{3,1} + (f, e_{2,2})e_{3,2} + ... + (f, e_{2,k})e_{3,k} + ... + ... = e_{1,1}.$$

By taking the inner product of both parts of the above relation with the element $e_{1,1}$ one obtains the contradictory fact $0 = 1$!)

Several other properties of V_1, V_2 can be found, the most important of which are summarized in the following propositions.

Proposition 2.4.1. The adjoint operators V_1^* and V_2^* of V_1 and V_2 are defined by:

$$V_1^* e_{i,j} = e_{i-1,j}, \quad i = 2, 3, ..., j = 1, 2, ..., \quad V_1^* e_{1,j} = 0, \quad j = 1, 2, ...$$

$$V_2^* e_{i,j} = e_{i,j-1}, \quad i = 1, 2, ..., j = 2, 3, ..., \quad V_2^* e_{i,1} = 0, \quad i = 1, 2, ...$$

Proof. For every $f \in H$ it is $(V_1 f, e_{1,j}) = 0$, when $j = 1, 2, ...$ or $(f, V^* e_{1,j}) = 0$. This last relation holds for every $f \in H$ and thus it will also hold for $f = V_1^* e_{1,j}$, which gives $\|V_1^* e_{1,j}\| = 0$ and as a consequence $V_1^* e_{1,j} = 0$, for $j = 1, 2, ...$ In the same way it is proved that $V_2^* e_{i,1} = 0$, for $i = 1, 2, ...$
Also, since

$$V_1 f = (f, e_{1,1}) e_{2,1} + (f, e_{1,2}) e_{2,2} + ... + (f, e_{1,k}) e_{2,k} + ... +$$

$$+ (f, e_{2,1}) e_{3,1} + (f, e_{2,2}) e_{3,2} + ... + (f, e_{2,k}) e_{3,k} + ... + ...$$

for $i \neq 1$, the inner product of both parts of the above relation with $e_{i,j}$ gives

$$(V_1 f, e_{i,j}) = (f, e_{i-1,j}) \Rightarrow (f, V_1^* e_{i,j}) = (f, e_{i-1,j}) \Rightarrow$$

$$\Rightarrow (f, V_1^* e_{i,j} - e_{i-1,j}) = 0, \forall f \Rightarrow V_1^* e_{i,j} = e_{i-1,j}.$$

In the same way it is proved that $V_2^* e_{i,j} = e_{i,j-1}$, for $j \neq 1$.   □

Proposition 2.4.2. The following relations hold between V_1, V_1^* and V_2, V_2^*:

$$V_1^* V_1 = I \text{ and } V_1 V_1^* = P_1, \quad V_2^* V_2 = I \text{ and } V_2 V_2^* = P_2,$$

where P_1, P_2 are projective operators and

$$\|V_1^*\| = \|V_1\| = 1, \quad \|V_2^*\| = \|V_2\| = 1.$$

Proof. It is left as exercise!   □

For the definitions of isometric, unitary adjoint and projective operators, one may consult any book in operator theory or functional analysis, see for example [35] or [36].

As already mentioned, the basic idea of the functional–analytic method presented in this paper, is the equivalent transformation of the PΔE under consideration to an operator equation in H or H_1, by using a specific isomorphism. This is accomplished with the following propositions:

Proposition 2.4.3. The mapping $\phi : H \rightarrow \ell_2$ defined by

$$\phi(f) = (f, e_{i,j}) = f(i, j) \tag{2.4.3}$$

is an isomorphism from H onto ℓ_2, i.e. is a linear, one by one mapping from H onto ℓ_2, which preserves the norm.

Proof. First of all the mapping defined by (2.4.3) is well-defined. Indeed, since $f \in H$ it is:

$$\|f(i, j)\|_{\ell_2}^2 = \sum_{i=1}^{\infty}\sum_{j=1}^{\infty} |f(i, j)|^2 = \sum_{i=1}^{\infty}\sum_{j=1}^{\infty} |(f, e_{i,j})|^2 = \|f\|^2 < +\infty.$$

By using the properties of the inner product, it is obvious that ϕ is linear. Also, ϕ is a one by one mapping onto ℓ_2. Indeed, if $f \in H$, $g \in H$ with $\phi(f) = \phi(g)$, then

$$(f - g, e_{i,j}) = 0 \Leftrightarrow f = g,$$

because $e_{i,j}$ is an orthonormal base of H.

Moreover, if $\alpha(i, j) \in \ell_2$, then there exists an $f \in H$ such that $\phi(f) = \alpha(i, j)$. This f is given by:

$$f = \sum_{i=1}^{\infty}\sum_{j=1}^{\infty} \alpha(i, j) e_{i,j},$$

and it belongs to H since

$$\|f\|^2 = \sum_{i=1}^{\infty}\sum_{j=1}^{\infty} |\alpha(i, j)|^2 = \|\alpha(i, j)\|_{\ell_2}^2 < +\infty.$$

Finally, the mapping ϕ preserves the norm since:

$$\|\phi(f)\|_{\ell_2}^2 = \sum_{i=1}^{\infty}\sum_{j=1}^{\infty} |f(i, j)|^2 = \sum_{i=1}^{\infty}\sum_{j=1}^{\infty} |(f, e_{i,j})|^2 = \|f\|^2.$$

$\square$

Proposition 2.4.4. The mapping $\phi : H_1 \rightarrow \ell_1$ defined by

$$\phi(f) = (f, e_{i,j}) = f(i, j) \tag{2.4.4}$$

is an isomorphism from H_1 onto ℓ_1.

Proof. The proof is similar to the proof of Proposition 2.4.3. $\square$

The element f defined by (2.4.3) or (2.4.4), is called the *abstract form* of $f(i, j)$. In general, if G is a mapping in ℓ_2 or ℓ_1 and N is a mapping in H or H_1, $N(f)$ is called the *abstract form* of $G(f(i, j))$ if

$$G(f(i, j)) = (N(f), e_{i,j}).$$

For almost all the linear terms appearing in PΔEs, the corresponding abstract forms have been found [30], as well as the abstract forms of many non–linear terms [33] (although the proofs are harder when it comes to non–linear terms). In the following, the abstract forms of various linear terms are given, as well as the abstract form of a non–linear term that will be needed in §6.

Proposition 2.4.5 (Linear Terms). The abstract form of

1. $f(i + p, j + q)$ is the element $(V_2^*)^q (V_1^*)^p f = (V_1^*)^p (V_2^*)^q f$,

2. $f(i - p, j - q)$ is the element $V_2^q V_1^p f = V_1^p V_2^q f$,

3. $f(i - p, j + q)$ is the element $(V_2^*)^q V_1^p f = V_1^p (V_2^*)^q f$,

4. $f(i + p, j - q)$ is the element $V_2^q (V_1^*)^p f = (V_1^*)^p V_2^q f$,

5. $a(i, j) f(i, j)$ is the element Af,

where A is the diagonal operator $Ae_{i,j} = a(i, j)e_{i,j}$ and $p, q \in \mathbb{N} \setminus \{0\}$.

Proof.   1. It is

$$f(i + p, j + q) = (f, e_{i+p, j+q}) = (f, V_1^p V_2^q e_{i,j}) =$$

$$= ((V_2^*)^q (V_1^*)^p f, e_{i,j}) = ((V_1^*)^p (V_2^*)^q f, e_{i,j}),$$

since V_1^*, V_2^* commute.

2. It is

$$f(i - p, j - q) = (f, e_{i-p, j-q}) = (f, (V_1^*)^p (V_2^*)^q e_{i,j}) =$$

$$= (V_2^q V_1^p u, e_{i,j}) = (V_1^p V_2^q u, e_{i,j}),$$

since V_1, V_2 commute.

3. It is

$$f(i - p, j + q) = (f, e_{i-p, j+q}) = (f, (V_1^*)^p V_2^q e_{i,j}) =$$

$$= ((V_2^*)^q V_1^p u, e_{i,j}) = (V_1^p (V_2^*)^q u, e_{i,j}),$$

since V_1, V_2^* commute.

4. It is

$$f(i+p, j-q) = (f, e_{i+p,j-q}) = (f, V_1^p(V_2^*)^q e_{i,j}) =$$

$$= (V_2^q(V_1^*)^p u, e_{i,j}) = ((V_1^*)^p V_2^q u, e_{i,j}),$$

since V_1^*, V_2 commute.

5. It is

$$\alpha(i,j)f(i,j) = \alpha(i,j)(f, e_{i,j}) = (f, \overline{\alpha(i,j)}e_{i,j}) = (f, A^* e_{i,j}) = (Af, e_{i,j}).$$

□

Proposition 2.4.6 (Non–Linear Term). The non-linear operator $N(f)$ of H_1 defined as follows:

$$N(f) = \sum_{i=1}^{\infty}\sum_{j=1}^{\infty}(f, e_{i,j})^2 e_{i,j} = \sum_{i=1}^{\infty}\sum_{j=1}^{\infty}[f(i,j)]^2 e_{i,j}. \tag{2.4.5}$$

is the abstract form of the non–linear term $[f(i,j)]^2$ in ℓ_1.

Proof. First of all the operator defined by (2.4.5) is well defined since

$$\|N(f)\|_1 = \sum_{i=1}^{\infty}\sum_{j=1}^{\infty}|(N(f), e_{i,j})| = \sum_{i=1}^{\infty}\sum_{j=1}^{\infty}|(f, e_{i,j})^2| \leq \|f\|_1 \sum_{i=1}^{\infty}\sum_{j=1}^{\infty}|f(i,j)| \Rightarrow$$

$$\Rightarrow \|N(f)\|_1 \leq \|f\|_1^2 < \infty,$$

since $f \in H_1$. Moreover,

$$(N(f), e_{i,j}) = \left(\sum_{i=1}^{\infty}\sum_{j=1}^{\infty}[f(i,j)]^2 e_{i,j}, e_{i,j}\right) = [f(i,j)]^2,$$

which implies that $N(f)$ is the abstract form of $[f(i,j)]^2$.   □

As it will be made clear in §6, in order to apply a specific useful fixed point theorem, the non–linear operator defined by (2.4.5) must be Frechét differentiable in a subspace of H_1. This is proved in the following:

Proposition 2.4.7 (Frechét Differentiability). The non-linear operator $N(f)$ defined by (2.4.5) is Frechét differentiable in H_1.

Proof. Since $N(f)$ is the abstract form of $[f(i,j)]^2$ and the spaces ℓ_1 and H_1 are isomorphic due to Proposition 2.4.4, it suffices to show that $[f(i,j)]^2$ is Frechét differentiable in ℓ_1. Indeed, the Frechét derivative of $[f(i,j)]^2$ at the point $f_0(i,j) \in \ell_1$ is the linear, bounded operator

$$D(f_0(i,j))f(i,j) = D_0 f = 2f_0(i,j)f(i,j).$$

The fact that D_0 is linear is obvious. The fact that D_0 is bounded follows easily since

$$\|D_0 f\|_{\ell_1} = \|2f_0(i,j)f(i,j)\|_{\ell_1} \leq 2\|f_0(i,j)\|_{\ell_1} \cdot \|f(i,j)\|_{\ell_1}.$$

Now D_0 is actually the Frechét derivative of $[f(i,j)]^2$ at the point $f_0(i,j) \in \ell_1$ since

$$\|[f_0(i,j) + h(i,j)]^2 - [f_0(i,j)]^2 - 2f_0(i,j)h(i,j)\|_{\ell_1} = \|[h(i,j)]^2\|_{\ell_1} \leq \|h(i,j)\|_{\ell_1}^2 \Rightarrow$$

$$\Rightarrow \frac{\|[f_0(i,j) + h(i,j)]^2 - [f_0(i,j)]^2 - 2f_0(i,j)h(i,j)\|_{\ell_1}}{\|h(i,j)\|_{\ell_1}} \leq \|h(i,j)\|_{\ell_1} \Rightarrow$$

$$\Rightarrow \lim_{\|h(i,j)\|_{\ell_1} \to 0} \frac{\|[f_0(i,j) + h(i,j)]^2 - [f_0(i,j)]^2 - 2f_0(i,j)h(i,j)\|_{\ell_1}}{\|h(i,j)\|_{\ell_1}} = 0.$$

$\square$

2.5   A linear example.

In this section, the method presented in §4 will be applied to the following linear initial value problem

$$f(i+1, j+1) = af(i,j) + g(i,j), \tag{2.5.1}$$

$$f(1,j), \;\; f(i,1) \text{ known complex sequences} \tag{2.5.2}$$

where $a \in \mathbb{C}$ and $g(i,j)$ is a known complex sequence. The question is to find conditions so that the initial value problem (2.5.1)-(2.5.2) has a unique solution in ℓ_2.

To begin with, the abstract forms of all the terms involved in (2.5.1) are needed. According to propositions 2.4.3 and 2.4.5 the abstract form of the unknown sequence $f(i,j)$ is an element $f \in H$ and the abstract form of the term $f(i+1, j+1)$ is the element $V_1^* V_2^* f = V_2^* V_1^* f \in H$. In order to obtain the abstract form, suppose $g \in H$ of the term $g(i,j)$, this sequence must belong to ℓ_2 (defined by (2.1.2)). This is the *fist assumption* of Theorem 2.5.2. Then, according to (2.4.3) one can write

$$(V_1^* V_2^* f, e_{i,j}) = a(f, e_{i,j}) + (g, e_{i,j}) \Rightarrow (V_1^* V_2^* f, e_{i,j}) - (af, e_{i,j}) - (g, e_{i,j}) = 0 \Rightarrow$$

$$\Rightarrow (V_1^* V_2^* f - af - g, e_{i,j}) = 0,$$

from which since $\{e_{i,j}\}$ is a base of H and the preceding relation holds for every $i, j = 1, 2, \ldots$ it follows

$$V_1^* V_2^* f - af - g = 0 \Rightarrow (V_1^* V_2^* - aI)f = g. \qquad (2.5.3)$$

Equation (2.5.3) is the equivalent to (2.5.1), operator equation in H. Taking into consideration the properties of V_1^* and V_2^* as stated in propositions 2.4.1 and 2.4.2, equation (2.5.3) becomes:

$$(V_2^* - aV_1)f = V_1 g + \sum_{j=1}^{\infty} c_j e_{1,j} \Rightarrow$$

$$\Rightarrow (I - aV_2 V_1)f = V_2 V_1 g + \sum_{j=1}^{\infty} c_j e_{1,j+1} + \sum_{i=1}^{\infty} d_i e_{i,1}, \qquad (2.5.4)$$

where the sequences c_j and d_i will be determined by the initial conditions (2.5.2) as follows:

By taking the inner product of both parts of (2.5.4) with the element $e_{1,1}$ one obtains:

$$(f, e_{1,1}) - a(V_2 V_1 f, e_{1,1}) = (V_2 V_1 g, e_{1,1}) + c_1(e_{1,2}, e_{1,1}) + c_2(e_{1,3}, e_{1,1}) + \ldots +$$

$$+ d_1(e_{1,1}, e_{1,1}) + d_2(e_{2,1}, e_{1,1}) + \ldots$$

from where since $\{e_{i,j}\}$ is an orthonormal base of H it follows:

$$(f, e_{1,1}) - a(V_1 f, V_2^* e_{1,1}) = (V_1 g, V_2^* e_{1,1}) + d_1 \overset{\textit{Proposition} \ 2.4.1}{\Rightarrow}$$

$$\Rightarrow (f, e_{1,1}) = d_1 \Rightarrow f(1,1) = d_1.$$

In the same way it can be proved, by taking the inner product of both parts of (2.5.4) with the element $e_{2,1}$, that $d_2 = f(2,1)$ and by induction that $d_i = f(i,1)$.

In a similar way, by taking the inner product of both parts of (2.5.4) with the element $e_{1,2}$ one obtains:

$$(f, e_{1,2}) - a(V_2 V_1 f, e_{1,2}) = (V_2 V_1 g, e_{1,2}) + c_1(e_{1,2}, e_{1,2}) + c_2(e_{1,3}, e_{1,2}) + \ldots +$$

$$+ d_1(e_{1,1}, e_{1,2}) + d_2(e_{2,1}, e_{1,2}) + \ldots$$

from where since $\{e_{i,j}\}$ is an orthonormal base of H it follows:

$$(f, e_{1,2}) - a(V_1 f, V_2^* e_{1,2}) = (V_1 g, V_2^* e_{1,2}) + c_1 \Rightarrow$$

$$\Rightarrow (f, e_{1,2}) - a(V_1 f, V_2^* e_{1,2}) = (V_1 g, V_2^* e_{1,2}) + c_1 \overset{Proposition\ 2.4.1}{\Rightarrow}$$

$$\Rightarrow (f, e_{1,2}) - a(V_1 f, e_{1,1}) = (V_1 g, e_{1,1}) + c_1 \Rightarrow$$

$$\Rightarrow (f, e_{1,2}) - a(f, V_1^* e_{1,1}) = (g, V_1^* e_{1,1}) + c_1 \overset{Proposition\ 2.4.1}{\Rightarrow}$$

$$\Rightarrow (f, e_{1,2}) = c_1 \Rightarrow f(1,2) = c_1.$$

In the same way it can be proved, by taking the inner product of both parts of (2.5.4) with the element $e_{1,3}$, that $c_2 = f(1,3)$ and by induction that $c_j = f(1, j+1)$.

Thus, equation (2.5.4) takes the form

$$(I - aV_2 V_1) f = V_2 V_1 g + \sum_{j=1}^{\infty} f(1, j+1) e_{1,j+1} + \sum_{i=1}^{\infty} f(i,1) e_{i,1}. \qquad (2.5.5)$$

Since the left–hand side part and the first term of the right–hand side of the preceding equation belong to H, in order (2.5.5) to be consistent, the last two terms of its right–hand side must also belong to H. This is guaranteed if $f(1,j)$ and $f(i,1)$ belong to the one–dimensional $\ell_2^{(1)}$ space. This is the *second assumption* of Theorem 2.5.2.

Equation (2.5.5) is the equivalent to the initial value problem (2.5.1)-(2.5.2), operator equation in H. Now one has to work with the abstract equation (2.5.5). Let $S = aV_2 V_1$. Then, $\|S\| \le |a|$. A well–known theorem in operator theory states that

Theorem 2.5.1. If T is a linear bounded operator of a Hilbert space H or a Banach space B, with $\|T\| < 1$, then $I - T$ is invertible with $\|(I-T)^{-1}\| \le \dfrac{1}{1 - \|T\|}$ and is defined on all H or B (see for example [35, p.70-71]).

Thus, if $|a| < 1$, then the operator $I - S$ has a unique inverse defined on all H. This is the *third assumption* of Theorem 2.5.2.

Then, from (2.5.5) one obtains

$$f = (I - S)^{-1} \left[V_2 V_1 g + \sum_{j=1}^{\infty} f(1, j+1) e_{1,j+1} + \sum_{i=1}^{\infty} f(i,1) e_{i,1} \right] \qquad (2.5.6)$$

and as a consequence, equation (2.5.5) has a unique solution in H given by (2.5.6). Equivalently, this means that under the above mentioned assumptions, the initial value problem (2.5.1)-(2.5.2) has a unique solution in ℓ_2. Moreover, from (2.5.6) one can obtain a bound of f. More precisely, the following holds:

$$\|f\| \le \|(I - S)^{-1}\| \left\| V_2 V_1 g + \sum_{j=1}^{\infty} f(1, j+1) e_{1,j+1} + \sum_{i=1}^{\infty} f(i,1) e_{i,1} \right\| \Rightarrow$$

$$\Rightarrow \|f\| \le \frac{1}{1-|a|} \left[\|g\| + \left\| \underbrace{\sum_{j=1}^{\infty} f(1,j+1)e_{1,j+1} + \sum_{i=1}^{\infty} f(i,1)e_{i,1}}_{h} \right\| \right],$$

where $h = \sum_{i=1}^{\infty} \sum_{j=1}^{\infty} h_{i,j} e_{i,j}$, with

$$h_{1,j+1} = c_j = f(1,j+1), \ \forall j$$

$$h_{i,j+1} = 0, \ \forall i \ne 1$$

$$h_{i,1} = d_i = f(i,1), \ \forall i$$

$$h_{i,j} = 0, \ \forall j \ne 1.$$

As a consequence

$$\|h\|^2 = \sum_{i=1}^{\infty} \sum_{j=1}^{\infty} |h_{i,j}|^2 = \sum_{j=1}^{\infty} |f(1,j+1)|^2 + \sum_{i=1}^{\infty} |f(i,1)|^2 \Rightarrow$$

$$\Rightarrow \|h\|^2 = \|f(1,j+1)\|^2_{\ell_2^{(1)}} + \|f(i,1)\|^2_{\ell_2^{(1)}}.$$

Thus

$$\|f\| \le \frac{1}{1-|a|} \left[\|g\| + \sqrt{\|f(1,j+1)\|^2_{\ell_2^{(1)}} + \|f(i,1)\|^2_{\ell_2^{(1)}}} \right].$$

Due to the fact that the used isomorphism preserves the norm (see Proposition 2.4.3) it follows that

$$\|f(i,j)\|_{\ell_2} = \|f\| \le \frac{1}{1-|a|} \left[\|g(i,j)\|_{\ell_2} + \sqrt{\|f(1,j+1)\|^2_{\ell_2^{(1)}} + \|f(i,1)\|^2_{\ell_2^{(1)}}} \right].$$

From the above it is obvious that the following theorem has been proved:

Theorem 2.5.2. Consider the initial value problem (2.5.1)-(2.5.2) and suppose that

(a) $g(i,j)$ belongs to the two–dimensional space ℓ_2 defined by (2.1.2),

(b) $f(1,j)$, $f(i,1)$ belong to the one–dimensional space $\ell_2^{(1)}$,

(c) $|a| < 1$.

Then, the initial value problem (2.5.1)-(2.5.2) has a unique solution in ℓ_2 satisfying

$$\|f(i,j)\|_{\ell_2} \le \frac{1}{1-|a|} \left[\|g(i,j)\|_{\ell_2} + \sqrt{\|f(1,j+1)\|^2_{\ell_2^{(1)}} + \|f(i,1)\|^2_{\ell_2^{(1)}}} \right] \qquad (2.5.7)$$

Remark 2.5.3. Suppose that $g(i,j) \equiv 0$, $f(1,j) \equiv 0$ and $f(i,1) \equiv 0$, for all $i,j = 1, 2, ...$ Then the initial value problem (2.5.1)-(2.5.2) has only the trivial (zero) solution, which of course belongs to ℓ_2.

Remark 2.5.4. Suppose that a is not a complex constant, but a complex sequence $a(i,j)$. In this case, it is $S = V_2 V_1 A$, where A is the diagonal operator $A e_{i,j} = a(i,j) e_{i,j}$, $i, j = 1, 2, ...$ and $\|S\| \leq \|A\| = \sup_{i,j} |a(i,j)|$. Then, conditions (a) and (b) remain the same, but condition (c) becomes

$$\sup_{i,j} |a(i,j)| < 1 \tag{2.5.8}$$

and relation (2.5.7) becomes

$$\|f(i,j)\|_{\ell_2} \leq \frac{1}{1 - \sup_{i,j} |a(i,j)|} \left[\|g(i,j)\|_{\ell_2} + \sqrt{\|f(1,j+1)\|_{\ell_2^{(1)}}^2 + \|f(i,1)\|_{\ell_2^{(1)}}^2} \right].$$

Remark 2.5.5. However, what if the sequence $a(i,j)$ doesn't satisfy (2.5.8)? In this case, one may assume another type of assumption for $a(i,j)$, namely that

$$\lim_{i,j \to \infty} a(i,j) = 0, \tag{2.5.9}$$

which can be considered as less restrictive than (2.5.8). In this case, the operator equation (2.5.5) becomes

$$(I - S)f = V_2 V_1 g + \sum_{j=1}^{\infty} f(1, j+1) e_{1,j+1} + \sum_{i=1}^{\infty} f(i,1) e_{i,1} \tag{2.5.10}$$

and the operator S is a compact operator, since A is a compact operator (due to (2.5.9)) and V_1, V_2 are bounded operators. Then the Fredholm alternative can be applied to (2.5.10). According to the Fredholm alternative (see for example [35, p. 243]), equation (2.5.10) has a unique solution in H if and only if the corresponding homogeneous equation

$$(I - S)f = 0 \Leftrightarrow (I - V_2 V_1 A)f = 0 \tag{2.5.11}$$

has only the trivial solution in H.

The solution of (2.5.11) is of the form

$$f = \sum_{i=1}^{\infty} \sum_{j=1}^{\infty} (f, e_{i,j}) e_{i,j} = \sum_{i=1}^{\infty} [(f, e_{i,1}) e_{i,1} + (f, e_{i,2}) e_{i,2} + ...] \tag{2.5.12}$$

and one can compute the coefficients $(f, e_{i,j})$ from (2.5.11), in the same way that the sequences c_j and d_i were computed in the proof of Theorem 2.5.2. More precisely, by taking the inner product of both parts of (2.5.11) with the element

- $e_{i,1}$, one obtains:

$$(f, e_{i,1}) - (V_2 V_1 A f, e_{i,1}) = 0 \Rightarrow (f, e_{i,1}) = 0, \ \forall \, i = 1, 2, \dots \qquad (2.5.13)$$

- $e_{i,2}$, one obtains:

$$(f, e_{i,2}) - (V_2 V_1 A f, e_{i,2}) = 0 \Rightarrow (f, e_{i,2}) - (V_1 A f, e_{i,1}) = 0 \Rightarrow$$

$$\Rightarrow \begin{cases} (f, e_{i,2}) = 0, & \text{for } i = 1 \\ (f, e_{i,2}) - (A f, e_{i-1,1}) = 0, & \text{for } i \geq 1 \end{cases} \Rightarrow$$

$$\Rightarrow \begin{cases} (f, e_{i,2}) = 0, & \text{for } i = 1 \\ (f, e_{i,2}) - \overline{a(i-1,1)}(f, e_{i-1,1}) = 0, & \text{for } i \geq 1 \end{cases} \Rightarrow$$

$$\overset{(2.5.13)}{\Rightarrow} \begin{cases} (f, e_{i,2}) = 0, & \text{for } i = 1 \\ (f, e_{i,2}) = 0, & \text{for } i \geq 1 \end{cases} \Rightarrow (f, e_{i,2}) = 0, \ \forall \, i = 1, 2, \dots \quad (2.5.14)$$

- $e_{i,3}$, one obtains:

$$(f, e_{i,3}) - (V_2 V_1 A f, e_{i,3}) = 0 \Rightarrow (f, e_{i,3}) - (V_1 A f, e_{i,2}) = 0 \Rightarrow$$

$$\Rightarrow \begin{cases} (f, e_{i,3}) = 0, & \text{for } i = 1 \\ (f, e_{i,3}) - (A f, e_{i-1,2}) = 0, & \text{for } i \geq 1 \end{cases} \Rightarrow$$

$$\Rightarrow \begin{cases} (f, e_{i,3}) = 0, & \text{for } i = 1 \\ (f, e_{i,3}) - \overline{a(i-1,2)}(f, e_{i-1,2}) = 0, & \text{for } i \geq 1 \end{cases} \Rightarrow$$

$$\overset{(2.5.14)}{\Rightarrow} \begin{cases} (f, e_{i,3}) = 0, & \text{for } i = 1 \\ (f, e_{i,3}) = 0, & \text{for } i \geq 1 \end{cases} \Rightarrow (f, e_{i,3}) = 0, \ \forall \, i = 1, 2, \dots \quad (2.5.15)$$

and by induction it follows that $(f, e_{i,j}) = 0$, for all $i, j = 1, 2, \dots$, which due to (2.5.12), implies that the solution of (2.5.11) is $f = 0$. As a consequence, the operator equation (2.5.10) has a unique solution in H. Then, conditions (a) and (b) of Theorem 2.5.2 remain the same, but condition (c) is substituted by condition (2.5.9) and relation (2.5.7) becomes

$$\|f(i,j)\|_{\ell_2} \leq \|(I - S)^{-1}\| \left[\|g(i,j)\|_{\ell_2} + \sqrt{\|f(1, j+1)\|^2_{\ell_2^{(1)}} + \|f(i,1)\|^2_{\ell_2^{(1)}}} \right].$$
$$(2.5.16)$$

It is obvious that although assumption (2.5.9) is more general than (2.5.8), the bound (2.5.16), is less concrete than (2.5.7) and has little practical interest, since the norm $\|(I - S)^{-1}\|$ or a bound of it, is not generally known or easy to find.

A concrete example

Consider the (real–valued) initial value problem

$$f(i+1, j+1) = \frac{1}{6} f(i,j), \qquad (2.5.17)$$

$$f(1,j) = \frac{1}{3^{j-1}}, \quad f(i,1) = \frac{1}{2^{i-1}}, \qquad (2.5.18)$$

which is of course of the form (2.5.1)-(2.5.2). The solution of this problem is the sequence $f(i,j) = \frac{1}{2^{i-1}3^{j-1}}$. This sequence belongs to ℓ_2, since

$$\sum_{i=1}^{\infty}\sum_{j=1}^{\infty} |f(i,j)|^2 = \sum_{i=1}^{\infty}\sum_{j=1}^{\infty} \left(\frac{1}{2^{i-1}3^{j-1}}\right)^2 = \sum_{i=1}^{\infty}\sum_{j=1}^{\infty} \left(\frac{6}{2^i 3^j}\right)^2 =$$

$$= 36 \sum_{i=1}^{\infty} \left(\frac{1}{4}\right)^i \sum_{j=1}^{\infty} \left(\frac{1}{9}\right)^j = 36 \left(\frac{1}{1-1/9} - 1\right)\left(\frac{1}{1-1/4} - 1\right) \Rightarrow$$

$$\Rightarrow \sum_{i=1}^{\infty}\sum_{j=1}^{\infty} |f(i,j)|^2 = \frac{3}{2} < \infty$$

and as a consequence $\|f(i,j)\|_{\ell_2} = \sqrt{1.5} \simeq 1.22474$.

Now, theorem 2.5.2 will be checked for the initial value problem (2.5.17)-(2.5.18) and it will be made clear, how this theorem "predicts" the behavior of the corresponding solution.

Assumptions (a) and (c) are obviously satisfied since $g(i,j) \equiv 0 \in \ell_2$ and $a = \frac{1}{6} < 1$. Assumption (b) is also satisfied since

$$\sum_{j=1}^{\infty} |f(1,j)|^2 = \sum_{j=1}^{\infty} \left(\frac{1}{3^{j-1}}\right)^2 = 9 \sum_{j=1}^{\infty} \left(\frac{1}{9}\right)^j = 9\left(\frac{1}{1-1/9} - 1\right) = \frac{9}{8} < \infty$$

and

$$\sum_{i=1}^{\infty} |f(i,1)|^2 = \sum_{i=1}^{\infty} \left(\frac{1}{2^{i-1}}\right)^2 = 4 \sum_{i=1}^{\infty} \left(\frac{1}{4}\right)^i = 4\left(\frac{1}{1-1/4} - 1\right) = \frac{4}{3} < \infty.$$

Thus, according to theorem 2.5.2, the initial value problem (2.5.17)-(2.5.18) has a unique solution in ℓ_2 satisfying

$$\|f(i,j)\|_{\ell_2} \leq \frac{1}{1-1/6}\left[\sqrt{\|f(1,j+1)\|_{\ell_2^{(1)}}^2 + \|f(i,1)\|_{\ell_2^{(1)}}^2}\right] \Leftrightarrow$$

$$\Leftrightarrow \|f(i,j)\|_{\ell_2} \leq \frac{6}{5}\left[\sqrt{\frac{1}{8} + \frac{4}{3}}\right] \simeq 1.44914,$$

a result which is in complete accordance with the "concrete" solution mentioned above.

2.6   A non-linear example.

In this section, the method presented in §4 will be applied to the following non–linear initial value problem

$$f(i+1, j+1) - a(i,j)f(i,j) = g(i,j) + [f(i,j)]^2 \qquad (2.6.1)$$

$$f(1,j), \ f(i,1) \ \text{known complex sequences} \qquad (2.6.2)$$

where $a(i,j)$ and $g(i,j)$ are known complex sequences. The question is to find conditions so that the initial value problem (2.6.1)-(2.6.2) has a unique solution in ℓ_1. As in §5, it can be found that the abstract form of (2.6.1)-(2.6.2) in the Banach space H_1 is

$$(I - S)f = V_2 V_1 g + \sum_{j=1}^{\infty} f(1, j+1)e_{1,j+1} + \sum_{i=1}^{\infty} f(i,1)e_{i,1} + V_2 V_1 N(f), \quad (2.6.3)$$

where $S = V_2 V_1 A$, A being the diagonal operator $Ae_{i,j} = a(i,j)e_{i,j}$, $i, j = 1, 2, \ldots$ and $N(f)$ defined by (2.4.5). If (and this is a big if!), the inverse of $I - S$ exists at least on a subset of H_1 and is bounded by a positive number L, then (2.6.3) takes the form

$$f = (I - S)^{-1}[V_2 V_1 g + \sum_{j=1}^{\infty} f(1, j+1)e_{1,j+1} + \sum_{i=1}^{\infty} f(i,1)e_{i,1} + V_2 V_1 N(f)] = \phi(f),$$

$$(2.6.4)$$

which is a non–linear operator equation and one may use several fixed point theorems (with the necessary assumptions of course), in order to prove that (2.6.4) has at least one or a unique solution in H_1.

Due to the fact that the non–linear operator $N(f)$ is Frechét differentiable, a fixed point theorem often used by the authors of this paper and Ifantis [27], is the following fixed point theorem of Earle and Hamilton [37]:

Theorem 2.6.1. If $f : X \to X$ is holomorphic, i.e. its Fréchet derivative exists, and $f(X)$ lies strictly inside X, then f has a unique fixed point in X, where X is a bounded, connected and open subset of a Banach space E. (By saying that a subset X' of X lies strictly inside X it is meant that there exists an $\epsilon_1 > 0$ such that $\|x' - y\| > \epsilon_1$ for all $x' \in X'$ and $y \in E - X$.)

In order to apply theorem 2.6.1, the simplest bounded, connected and open subset of H_1 is considered, namely the open ball $B(0, R) = \{f \in H_1 : \|f\|_1 < R\}$, $R > 0$. However, it is noted again, that any fixed point theorem can be properly applied to (2.6.4).

Returning to (2.6.4), suppose that $f \in B(0, R)$. Then $\|f\|_1 < R$ and

$$\|\phi(f)\|_1 \le L[\|g\|_1 + \|\sum_{j=1}^{\infty} f(1, j+1)e_{1,j+1} + \sum_{i=1}^{\infty} f(i,1)e_{i,1}\|_1 + \|N(f)\|_1] \Rightarrow$$

$$\Rightarrow \|\phi(f)\|_1 \le L \left(\|g\|_1 + \|h\|_1 + \|f\|_1^2 \right) \Rightarrow$$
$$\Rightarrow \|\phi(f)\|_1 \le L \left(\|g\|_1 + \|h\|_1 \right) + LR^2, \tag{2.6.5}$$

where the element h is defined as in §5.

Define the function $P(R) = R - LR^2$, which attains its maximum $P_0 = \dfrac{1}{4L}$ at the point $R_0 = \dfrac{1}{2L}$. Then for $\|f\|_1 \le R_0 - \epsilon < R_0$, it follows that if

$$L \left(\|g\|_1 + \|h\|_1 \right) \le P_0 - \epsilon < P_0$$

or

$$4L^2 \left(\|g\|_1 + \|h\|_1 \right) < 1, \tag{2.6.6}$$

then (2.6.5) gives

$$\|\phi(f)\|_1 \le P_0 - \epsilon + LR_0^2 = R_0 - LR_0^2 - \epsilon + LR_0^2 = R_0 - \epsilon < R_0,$$

which means that theorem 2.6.1 is applied to (2.6.4) and as a consequence if (2.6.6) holds, then (2.6.4) has a unique solution in H_1 bounded by R_0.

From the above it is obvious that the following theorems have been proved:

Theorem 2.6.2. Consider the initial value problem (2.6.1)-(2.6.2) and suppose that
(a) $g(i,j)$ belongs to the two–dimensional space ℓ_1 defined by (2.1.1),
(b) $f(1,j)$, $f(i,1)$ belong to the one–dimensional space $\ell^{(1)}$,
(c) $\lim\limits_{i,j \to \infty} a(i,j) = 0$.
Then, there exists an $L > 0$ such that if

$$4L^2 \left(\|g(i,j)\|_{\ell_1} + \|f(1,j+1)\|_{\ell_1^{(1)}} + \|f(i,1)\|_{\ell_1^{(1)}} \right) < 1, \tag{2.6.7}$$

the initial value problem (2.6.1)-(2.6.2) has a unique solution in ℓ_1, bounded by $R_0 = \dfrac{1}{2L}$.

Theorem 2.6.3. Consider the initial value problem (2.6.1)-(2.6.2) and suppose that
(a) $g(i,j)$ belongs to the two–dimensional space ℓ_1 defined by (2.1.1),
(b) $f(1,j)$, $f(i,1)$ belong to the one–dimensional space $\ell^{(1)}$,
(c) $\sup\limits_{i,j} |a(i,j)| < 1$.
Then, if

$$4 \left(\|g(i,j)\|_{\ell_1} + \|f(1,j+1)\|_{\ell_1^{(1)}} + \|f(i,1)\|_{\ell_1^{(1)}} \right) < \left(1 - \sup\limits_{i,j} |a(i,j)| \right)^2, \tag{2.6.8}$$

the initial value problem (2.6.1)-(2.6.2) has a unique solution in ℓ_1, bounded by $R_0 = \dfrac{1 - \sup_{i,j} |a(i,j)|}{2}$.

Theorem 2.6.4. Consider the initial value problem (2.6.1)-(2.6.2), where the sequence $a(i, j)$ is replaced by the constant complex number a and suppose that
(a) $g(i, j)$ belongs to the two–dimensional space ℓ_1 defined by (2.1.1),
(b) $f(1, j)$, $f(i, 1)$ belong to the one–dimensional space $\ell^{(1)}$,
(c) $|a| < 1$.
　　Then, if

$$4 \left(\|g(i, j)\|_{\ell_1} + \|f(1, j + 1)\|_{\ell_1^{(1)}} + \|f(i, 1)\|_{\ell_1^{(1)}} \right) < (1 - |a|)^2, \qquad (2.6.9)$$

the corresponding initial value problem has a unique solution in ℓ_1, bounded by
$R_0 = \dfrac{1 - |a|}{2}$.

Remark 2.6.5. Relations (2.6.6), (2.6.7), (2.6.8) and (2.6.9) depend only on the initial conditions and the parameters appearing in the PΔE under consideration. These inequalities define a region where there exists a unique solution in ℓ_1 for the problem under consideration. As already mentioned in §3, this solution tends to zero as i, j tend to infinity, which means that zero is a locally asymptotically stable point of the PΔE under consideration and this is the reason, for which the above mentioned inequalities are also called *regions of attraction*.

2.7　Relevant material.

　　Some relevant material regarding ordinary difference equations, as well as some future thoughts concerning the numerical solution of PDEs is given.

- A similar (in philosophy) approach for the study of ordinary difference equations has recently been employed in (EP) and (P), where the crucial observation was that "An initial value problem for a difference equation is equivalent to a fixed point problem in a sequence space" (not necessarily equipped with an inner product).

　(EP) K. Ey, and C. Pötzsche, "Asymptotic behavior of recursions via fixed point theory", J. Math. Anal. Appl. Vol. 337 pp. 1125–1141, 2008.

　(P) C. Pötzsche, "A functional-analytic approach to the asymptotics of recursion", Proc. Amer. Math. Soc. Vol. 137, No. 10 pp. 3297–3307, 2009.

- Recently, the method presented in §4, but for ordinary difference equations (OΔEs), was combined with a similar (in philosophy) method for ordinary differential equations (ODEs), which was developed by Ifantis in (I78), (I87a) and (I87b) and used by the authors in a series of papers. By this combination it is possible to obtain the discrete equivalent of an ODE. In this way, one may solve numerically initial or boundary value problems in $\mathbb{R}$ or $\mathbb{C}$, with a method which guarantees not only the uniqueness of the solution, but also the convergence of the approximate solution to the true solution of the ODE (see (PSTa), (PSTb)). It is hoped that an analogue combination of methods could be used in the future for the numerical solution of PDEs.

(I78) E. K. Ifantis, "An existence theory for functional-differential equations and functional-differential systems", J. Differential Equations., Vol. 29 pp. 86–104, 1978.

(I87a) E. K. Ifantis, "Analytic solutions for nonlinear differential equations", J. Math. Anal. Appl., Vol. 124 pp. 339–380, 1987.

(I87b) E. K. Ifantis, "Global analytic solutions of the radial nonlinear wave equation", J. Math. Anal. Appl., Vol. 124 pp. 381–410, 1987.

(PSTa) E. N. Petropoulou, P. D. Siafarikas and E. E. Tzirtzilakis, "A "discretization" technique for the solution of ODEs", J. Math. Anal. Appl., Vol. 331 pp. 279–296, 2007.

(PSTb) E. N. Petropoulou, P. D. Siafarikas and E. E. Tzirtzilakis, "A "discretization" technique for the solution of ODEs II", Numer. Funct. Anal. Optim., Vol. 30 (No. 5-6) pp. 613–631, 2009.

References

[1] R.P. Agarwal. *Difference equations and inequalities. Theory, methods and applications.* Marcel Dekker, 1992.

[2] S.S. Cheng. *Partial difference equations*, volume 3 of *Advances in Discrete Mathematics and Applications.* Taylor & Francis, 2003.

[3] W. G. Kelley and A. C. Peterson. *Difference equations. An introduction with applications.* Academic Press, 1991.

[4] R. E. Mickens. *Difference equations. Theory and applications.* Van Nostrand Reinhold Co., 1990.

[5] B. Zhang and Y. Zhou. *Qualitative Analysis of Delay Partial Difference Equations*, volume 4 of *Contemporary Mathematics and Its Applications.* Hindawi Publishing Corporation, New York, USA, 2007.

[6] R. E. Mickens. *Nonstandard finite difference models of differential equations.* World Scientific, 1993.

[7] K. W. Morton. Functional analysis and the numerical solution of partial differential equations. *IMA Bulletin (Applied Functional Analysis in Teaching and Research)*, 10(4):108–111, 1974.

[8] E. E. Tzirtzilakis and N. G. Kafoussias. Numerical schemes and difference equations. *Some Recent Advances in Partial Difference Equations (this e-book)*, pages 111–140, Bentham Science Publishers, 2010.

[9] P. D. Lax. Numerical solution of partial differential equations. *Amer. Math. Monthly*, 72(2):74–84, 1965.

[10] A. N. Sharkovsky, Yu. L. Maistrenko, and E. Yu Romanenko. *Difference equations and their applications*. Kluwer Academic Publishers, 1993.

[11] E. N. Petropoulou and P. D. Siafarikas. A functional-analytic method for the study of difference equations. *Adv. Differ. Equat.*, 2004:3:237–248, 2004.

[12] B. Noble. Introduction: Function spaces. *IMA Bulletin (Applied Functional Analysis in Teaching and Research)*, 10(4):98–103, 1974.

[13] L. B. Rall. Applied functional analysis in teaching and research–some afterthoughts. *IMA Bulletin (Applied Functional Analysis in Teaching and Research)*, 10(4):126–127, 1974.

[14] S. S. Cheng and R. Medina. Bounded and positive solutions of discrete steady state equations. *Tamkang J. Math.*, 31(2):131–135, 2000.

[15] P.J.Y. Wong. Triple positive solutions of conjugate boundary value problems ii. *Comput. Math. Appl.*, 40:537–557, 2000.

[16] P.J.Y. Wong and R. P. Agarwal. Existence of multiple positive solutions of discrete two–point right focal boundary value problems. *J. Differ. Equations Appl.*, 5:517–540, 1999.

[17] P.J.Y. Wong and R. P. Agarwal. Multiple solutions of difference and partial difference equations with lidstone conditions. *Math. Comput. Modelling*, 32:699–725, 2000.

[18] P.J.Y. Wong and R. P. Agarwal. Criteria for multiple solutions of difference and partial difference equations subject to multipoint conjugate conditions. *Nonlinear Anal.*, 40:629–661, 2000.

[19] P.J.Y. Wong and L. Xie. Three symmetric solutions of lidstone boundary value problems for difference and partial difference equations. *Comput. Math. Appl.*, 45:1445–1460, 2003.

[20] H. Amann. Fixed point equations and nonlinear eigenvalue problems in ordered banach spaces. *SIAM Rev.*, 18:620–709, 1976.

[21] D. Guo and V. Lakshikantham. *Nonlinear problems in abstract cones*. Academic Press, 1988.

[22] R. W. Leggett and L. R. Williams. Multiple positive fixed points of nonlinear operators on ordered banach spaces. *Indiana Univ. Math. J.*, 28:673–688, 1979.

[23] V. Volterra. *Theory of functionals and of integral and integro-differential equations*. Dover publications, 1959.

[24] S. Bergman. Linear operators in the theory of partial differential equations. *Trans. Amer. Math. Soc.*, 53(1):130–155, 1943.

[25] A. H. Nayfeh. *Introduction to Perturbation Techniques*. Wiley Interscience Publication, 1993.

[26] E. K. Ifantis. Spectral theory of the difference equation $f(n+1)+f(n-1) = [e - \phi(n)]f(n)$. *J. Math. Phys.*, 10(3):421–425, 1969.

[27] E. K. Ifantis. On the convergence of power-series whose coefficients satisfy a poincaré-type linear and nonlinear difference equation. *Complex Variables*, 9:63–80, 1987.

[28] E. N. Petropoulou and P. D. Siafarikas. Bounded solutions and asymptotic stability of nonlinear difference equations in the complex plane. *Arch. Math. (Brno)*, 36(2):139–158, 2000.

[29] E. N. Petropoulou. On some specific non-linear ordinary difference equations. *Arch. Math. (Brno) (CDDE Issue)*, 36:549–562, 2000.

[30] E. N. Petropoulou and P. D. Siafarikas. Bounded solutions and asymptotic stability of nonlinear difference equations in the complex plane ii. *Comput. Math. Appl. (Special Issue: Advances in Difference Equations III)*, 42(3–5):427–452, 2001.

[31] E. N. Petropoulou and P. D. Siafarikas. Existence of complex ℓ_2 solutions of linear delay systems of difference equations. *J. Differ. Equat. Appl.*, 11(1):49–62, 2005.

[32] E. N. Petropoulou and P. D. Siafarikas. Bounded solutions of a class of linear delay and advanced partial difference equations. *Dynam. Systems Appl.*, 10(2):243–260, 2001.

[33] E. N. Petropoulou and P. D. Siafarikas. Solutions of non-linear delay and advanced partial difference equations in the space $\ell^1_{\times \times \times}$. *Comput. Math. Appl. (Special Issue: Advances in Difference Equations IV)*, 45:905–934, 2003.

[34] E. N. Petropoulou. On the eigenvalue problem of a class of linear partial difference equations. *J. Differ. Equat. Appl.*, January 2010.

[35] I. Gohberg and S. Goldberg. *Basic operator theory*. Birkhäuser, 1980.

[36] K. Maurin. *Methods of Hilbert spaces*. Polish Scientific Publishers, 1972.

[37] C. J. Earle and R. S. Hamilton. A fixed point theorem for holomorphic mappings. *Global Analysis (Proc. Sympos. Pure Math., Berkeley, California 1968*, pages 61–65.

CHAPTER 3

Partial difference equations and their application in systems theory

Jiří Gregor and Josef Hekrdla

Department of Mathematics, Faculty of Electrical Engineering, Czech Technical University, Technická 2, 166 27 Praha 6, Czech Republic

e-mails: gregorj@math.feld.cvut.cz, hekrdla@math.feld.cvut.cz

Abstract: This paper is a survey of basic results of the theory of partial difference equations (PDE's) and its application in multidimensional systems theory. Existence and uniqueness theorems of solutions of initial value problems, some boundary value problems, fundamental solutions for linear PDE's are presented. Most results are extended to systems of linear PDE's. Recursive solutions play an important role not only in these theorems but can also be used to find growth estimates and formulate further qualitative properties of PDE's. Application of these results to input–output relations of linear multidimensional systems (called also nD–systems) enables to introduce concepts analogous to time–invariance, causality, weight functions, impulse response and similar ones, well known from (one–dimensional) systems theory. In this paper fundamental results concerning some PDE are described. Some generalizations of earlier published results are introduced.

Key words and phrases: Difference equations, partial difference equations, n–dimensional sequences, initial value problems, boundary value problems, recursive solution, system theory, discrete system, shift invariant system, translation invariant system, BIBO stability, initial state stable system, input stable system.

Eugenia N. Petropoulou (Ed)

3.1 Introduction.

Difference equations are functional equations where the unknown functions are defined on a countable set. In mostly considered cases this set is a finite or infinite interval of integers and the unknowns are sequences. Commonly, for functions defined on such set $A \subseteq \mathbb{Z}$ ($\mathbb{Z}$ – the set of all integers), these equations for an unknown sequences y can be described generally as $F(\alpha, y(\alpha + \beta_1), \ldots, y(\alpha + \beta_m)) = 0$, where $\alpha \in A$ and $B = \{\beta_1, \ldots, \beta_m\}$ is a finite subset of $\mathbb{Z}$. If F is a linear form, for example $F(\alpha, y(\alpha + \beta_1), \ldots, y(\alpha + \beta_m)) = \sum_{\beta \in B} a(\alpha, \beta) y(\alpha + \beta) - x(\alpha)$, such equations are called (ordinary) linear difference equations and the difference $\max(B) - \min(B)$ is its order. Moreover, if the right–hand side x is identically zero, the equation is called homogeneous.

Replacing $\mathbb{Z}$ everywhere in this description with $\mathbb{Z}^n$ (Cartesian power of $\mathbb{Z}$) we obtain a formal description of partial difference equations with only one, seemingly unimportant, shortcoming: How can the order of such equation be defined? We want to show that attempts to answer this "simple" question reveals most of the fundamental distinctions between ordinary and partial difference equations.

Perhaps it is in order to explain why these equations are called "difference equations". For sequences the concept of differences can be introduced and some of their properties resemble analogous properties of derivatives. This way the concept of difference equations can be considered as analogous to the concept of differential equations. A number of methods to solve differential equations have their analogues in difference equations. This is true also for partial differences and of differential equations. Many authors pursue this way and bring interesting examples and some results on PDE's (*Partial Difference Equations*), (see Agarwal [1], Kelly–Peterson and others). In this paper we will not use the symbolics of partial differences like $\triangle_{ijk}$ since it seems to be rather restrictive. Each partial difference fit our notation, i.e. $\triangle_i y(\alpha) = y(\alpha + e_i) - y(\alpha)$, $e_i = (0, \ldots, 1, \ldots, 0)$ has 1 on i–th position, but difference equations $F(\alpha, y(\alpha + \beta_1), \ldots, y(\alpha + \beta_m)) = 0$ not always be rewrite into the form where partial differences are used only.

While the theory of one–dimensional difference equations is well developed, there are substantial distinctions between the one–dimensional and multi–dimensional cases. At this stage two of them have to be mentioned.

Any subset A of integers $\mathbb{Z}$ is a naturally ordered set by common ordering $\leq$ and if A is bounded from below, it becomes naturally well–ordered set. It contrasts with $A \subseteq \mathbb{Z}^n$, $n \geq 2$, where there is no such natural ordering.

In case of ordinary linear homogeneous equations the space of solutions is a finite dimensional linear space, while the dimension of the space of solutions of partial linear homogeneous equations is not of finite dimension. Moreover, one should be careful using the notion of "dimension". Since there are one–to–one correspondences between $\mathbb{Z}^m$, $\mathbb{Z}^n$ for $m \neq n$. This fact is effectively used in many environments, see [2], for example.

The history of investigations of PDE's goes back for about half a century, although some examples are much older. The simplest ones are the equation satisfied by combinatorial numbers $\begin{pmatrix} m \\ n \end{pmatrix}$ or those defining Stirling numbers, Euler numbers etc. Many other examples have been investigated, mostly those resulting from discretization of partial differential equations.

For a systematic approach to PDE the basic results must include theorems on existence and uniqueness of solution. Such theorems have to be formulated in terms of the sets A, B and the functions F, x. A systematic approach to linear PDE's with constant coefficients have been given in [3].

A few years ago one of the first monographs appeared containing a comprehensive bibliography.

Substantial efforts have been given to applications of PDE's in theory of multidimensional signals and systems theory (see e. g. [4, 5], [6] and many others as well as hundreds of papers in journals and conferences). A specialized journal called "Multidimensional Systems and Signal Processing" edited by N. K. Bose is now in its 20 years of existence.

The aim of this paper is to give a short survey of basic results on existence and unicity of solutions of PDE's under various settings (including systems of PDE's but concentrating on linear equations), some results on qualitative properties of solutions of PDE's such as growth estimates, and some applications of PDE's in system theory, mainly with systems considered as input–output relations.

We will now summarize the basic notation of this paper.

Capitals will denote sets, mostly subsets of $\mathbb{Z}^n$. Mappings defined on subsets of $\mathbb{Z}^n$ with values in $\mathbb{C}$ i. e. mappings like $x : A \rightarrow \mathbb{C}$, $A \subseteq \mathbb{Z}^n$ will be called sequences (or discrete functions) and will be denoted by letters x, y, z ..., f, g, h .... The set of all such sequences will be denoted as $\mathbb{C}^A$. In $\mathbb{Z}^n$ the elements are denoted as Greek letters $\alpha, \beta, \gamma, \ldots$ with it's components $\alpha = (\alpha^1, \alpha^2, \ldots, \alpha^n)$, sum of two such elements is defined component–wise and often the "shift" of sets defined by

$$A + \beta = \{\gamma : \gamma = \alpha + \beta, \alpha \in A\},$$

$$A + B = \{\gamma : \gamma = \alpha + \beta, \alpha \in A, \beta \in B\}.$$

Greek letters also denote integer valued mappings like $\sigma : A \rightarrow X$, $A \subseteq \mathbb{Z}^n$, $X \subseteq \mathbb{Z}^m$, $n \geq 2$, $m \geq 1$.

It will turn out in the sequel that we should make use of a special concept of set theory, namely the concept of ordering.

Definition 3.1.1. Let A be a set. A pair $(A, \preccurlyeq)$ is called an *ordered set* if a binary relation $\preccurlyeq$ satisfies the following properties for any elements $a, b, c \in A$.

- $a \preccurlyeq a$ for any $a \in A$,
  we say that the relation $\preccurlyeq$ is *reflective* on A.

- If $a \preccurlyeq b$ and $b \preccurlyeq c$ then $a \preccurlyeq c$,
  we say that the relation $\preccurlyeq$ is *transitive* on A.

- If $a \preccurlyeq b$ and $b \preccurlyeq a$ then $a = b$,
  We say that the relation $\preccurlyeq$ is *antisymmetrical* on A.

Commonly, the relation $a \prec b$ means $a \preccurlyeq b$ and $a \neq b$, similarly we suppose that $b \succcurlyeq a$ is equivalent to $a \preccurlyeq b$ and the like. Moreover, we will frequently used more rich ordering structures.

Definition 3.1.2. An ordered set $(A, \preccurlyeq)$ is called a *linear ordered set* if any two elements of A are comparable, so for every $a, b \in A$ either $a \preccurlyeq b$ or $b \preccurlyeq a$.

An ordered set $(A, \preccurlyeq)$ is called a *well ordered set*, if any of its nonempty subset has the least element, so if $\emptyset \neq X \subseteq A$ then there is an element $a \in X$ for which $a \preccurlyeq x$ for all $x \in X$.

It should be mentioned that a well ordered set is also a linear ordered set. Note that the set $\mathbb{Z}$ is linearly ordered by common ordering relation $\leq$, but it is not well–ordered. However, any subset $A \subseteq \mathbb{Z}$ bounded from below is well ordered, so the set of natural numbers $\mathbb{N} = n : n \in \mathbb{Z}, n \geq 0$ is well ordered by $\leq$.

3.2   Solutions of PDE.

Let's define a PDE more precisely:

Definition 3.2.1. The equation

$$F(\alpha, y(\alpha + \beta_1), \ldots, y(\alpha + \beta_m)) = 0, \tag{3.2.1}$$

where $\alpha \in A \subseteq \mathbb{Z}^n$, $n \geq 2$, $B = \{\beta_1, \beta_2, \cdots, \beta_m\} \subseteq \mathbb{Z}^n$, represents a *partial difference equation*. The number of elements B we will denote $|B|$ and we always suppose $|B| \geq 2$. If the equation above can be written in the form

$$\sum_{\beta \in B} a(\alpha, \beta) y(\alpha + \beta) = x(\alpha), \tag{3.2.2}$$

where $a : A \times B \to \mathbb{C}$, $y : A + B \to \mathbb{C}$, $x : A \to \mathbb{C}$ we will call it *linear*. This equation is called homogeneous if $x \equiv 0$. If we define a linear operator $\mathcal{L}[y](\alpha) = \sum_{\beta \in B} a(\alpha, \beta) y(\alpha + \beta)$, we can write (3.2.2) in a short form

$$\mathcal{L}[y] = x. \tag{3.2.3}$$

Any mapping $y : A + B \to \mathbb{C}$ satisfying equation (3.2.1) is called its *solution*.

To illustrate the previous definition and the basic concepts let us consider the following PDE's with nonconstant coefficients the solution of which are the known Stirling numbers[1].[2] (In some examples a more transparent and self–explanatory notation is used for elements of $\mathbb{Z}^2$.)

---

[21]Stirling numbers of the first kind are coefficients in expansion of $n! \begin{pmatrix} m \\ n \end{pmatrix}$ in powers of m. Stirling numbers of the first and second kind also appear in some statistic and combinatorial calculations.

Example 3.2.2. Stirling numbers of the first and second kind can be defined as certain solutions of the following equations on a triangular shaped region. Equation

$$y(m,n) = (1-m)y(m-1,n) + y(m-1,n-1) \tag{3.2.4}$$

and equation

$$y(m,n) = ny(m-1,n) + y(m-1,n-1) \tag{3.2.5}$$

for $1 \leq m$, $1 \leq n \leq m$, with prescribed initial values on the boundary $y(0,0) = 1$, $y(k,k+1) = 0$ for $k \geq 0$ and $y(m,0) = 0$ for $m \geq 1$, define the Stirling numbers of the first and second kind, respectively.

These equations are of the form (3.2.2), homogeneous and with nonconstant coefficients, $x(m,n) = 0$ on the region $A = \{(m,n) : (m,n) \in \mathbb{Z}^2, 1 \leq m, 1 \leq n \leq m\}$. The set B (frequently called mask) consists of three elements $B = \{\beta_1, \beta_2, \beta_3\}$, where $\beta_1 = (0,0)$, $\beta_2 = (-1,0)$, $\beta_3 = (-1,-1)$. The solution y is defined on the region $A+B$ and (boundary) initial values defined on the set $(A+B) \smallsetminus A$. When these boundary initial values are predefined, the values of y in both cases can be recursively calculated on the set A by the equations given above. The solution y is uniquely determined by this recursion. In the next Figure 3.1 we illustrate the situation.

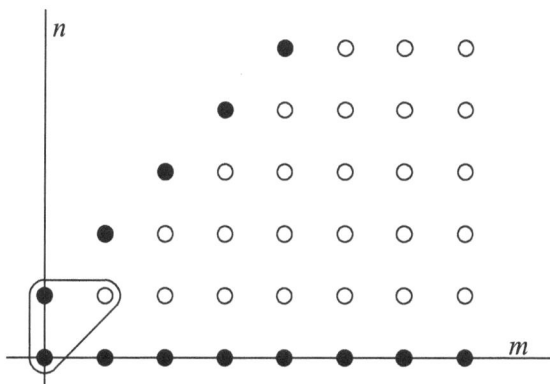

Figure 3.1: Grid of initial values and mask used in calculation of Stirling numbers. Black circles are points where initial values are predefined (they form the boundary of the region $A+B$), white points denote an area A where a solution y can be recursively calculated.

The rounded triangle denotes the shifted mask $(1,1)+B$ in the position, where the first "unknown" value $y(1,1)$ can be calculated. This is the value which has to be calculated first. Having this new value $y(1,1)$ we can continue to calculate $y(2,1)$ or $y(2,2)$.

The *leading* element β of the mask B, (which might have to be chosen differently for various elements $\alpha \in A$, we will denote it β_α), is that for which the new value

$y(\alpha + \beta_\alpha)$ is calculated from the following equation

$$y(\alpha + \beta_\alpha) = \frac{1}{a(\alpha, \beta_\alpha)} \left(x(\alpha) - \sum_{\beta \in B_\alpha} a(\alpha, \beta) y(\alpha + \beta) \right). \qquad (3.2.6)$$

The corresponding coefficient $a(\alpha, \beta_\alpha)$ is called a leading coefficient. In the equation above we denote $B_\alpha = B \setminus \{\beta_\alpha\}$.

We observe that the values of y can be recursively calculated in some order, using only the initial values and the values of the solution y found in previous steps. This ordering plays a significant role not only in recursive calculations of y but it is an inherent part of basic questions of existence and uniqueness of such solutions. In the next section we show that for linear PDE it can always be found (under very mild sufficient conditions) an appropriate ordering and a set of initial values under which the solution of (3.2.2) will exist and will be unique.

3.2.1   Initial value problem.

In this section we will consider linear PDE's, but some corollaries for non – linear PDE's will also be formulated. It will turn out that the existence and uniqueness of solution is based on the possibility of its recurrent evaluation. The problem of existence and uniqueness of solutions is of fundamental importance. The following theorem solves this problem for linear PDE's and provides some insight into recursive calculation of the solution.

Theorem 3.2.3. Let us be given a linear difference equation (3.2.2) with nonzero coefficients; let A, B be nonempty subsets of $\mathbb{Z}^n$, B finite and with at least two elements, formally: $a : A \times B \to \mathbb{C} \setminus \{0\}$, $A, B \subseteq \mathbb{Z}^n$, $A \neq \emptyset$, $2 \leq |B| < \infty$.

Then, there exists a set G, $G \subseteq A + B$, and a well ordering $\preccurlyeq$ of $A + B$, such that:

1. Any function $g : G \to \mathbb{C}$ can be extended to a unique solution $y : A + B \to \mathbb{C}$ of (3.2.2),

2. any value $y(\alpha)$, $\alpha \in A + B$, of the solution can be calculated from (3.2.2) by a finite number of arithmetic operations using only values $y(\alpha')$ for $\alpha' \prec \alpha$.

For the proof see [7, 8].

Theorem 3.2.3 is a starting point of PDE's theory and it is also important for construction of recursive algorithms solving (3.2.2).

The proof of the Theorem 3.2.3 is based on the following facts. For any finite subset $B \subseteq \mathbb{Z}^n$ with at least two elements, an integer valued function $\sigma : \mathbb{Z}^n \to \mathbb{Z}$ can always be constructed which is one – to – one on the set B. The function is linear and it is determined by some integers $N_i \in \mathbb{Z}$, $\sigma(\alpha) = N_1 \alpha^1 + N_1 \alpha^2 + \ldots + N_n \alpha^n$. Geometrically, the function σ projects the "integer space $\mathbb{Z}^n$" into the line the direction of which is determined by the integer valued vector $(N_1, \cdots, N_n)$. The hyperplane, orthogonal to this vector containing the origin of $\mathbb{Z}^n$, divides the space $\mathbb{Z}^n$ (and any of its subset

$A \subseteq \mathbb{Z}^n$) into two parts, $A_0 = A \cap \{\alpha : \sigma(\alpha) \geq 0\}$ and $A_1 = A \cap \{\alpha : \sigma(\alpha) < 0\}$. For these subsets we chose different leading elements of B. If $\alpha \in A_0$ we define β_α such that $\sigma(\beta_\alpha) = max\,(\sigma(B))$, for $\alpha \in A_1$ we chose β_α such that $\sigma(\beta_\alpha) = min\,(\sigma(B))$ (this is why we need at least two elements in B). Moreover, let's define a function $\rho : A \to A + B$ determined by the relation $\alpha \mapsto \alpha + \beta_\alpha$. This function shifts subsets A_0 and A_1 in the directions of the corresponding leading elements, respectively. The function ρ is one-to-one. Finally, we define the set $G = (A + B) \smallsetminus \rho(A)$ on which the initial values can arbitrary be prescribed. The Theorem 3.2.3 shows that a well ordering $\preccurlyeq$ of $\rho(A) \subseteq A + B$ can always be found and that any value $y(\alpha)$ for $\alpha \in \rho(A)$ can recursively be calculated by (3.2.6) using only the values prescribed on G or the values $y(\alpha')$, $\alpha' \in \rho(A)$, calculated in the previous steps for which $\alpha' \prec \alpha$.

To explain the main ideas of the proof of 3.2.3 we will step by step construct the solutions of equations (3.2.4), (3.2.5).

Example 3.2.4. Reconsider Example 3.2.2 now with $A = \mathbb{Z}^2$. We see that a function $\sigma : \mathbb{Z}^2 \to \mathbb{Z}, \sigma(\alpha^1, \alpha^2) = \alpha^1 + \alpha^2$, projects the mask $B = \{\beta_1, \beta_2, \beta_3\}$ on the line directed by the vector $(1, 1)$ injectively and $\sigma(\beta_1) = max\,(\sigma(B)) = 0$, $\sigma(\beta_3) = min\,(\sigma(B)) = -2$. So, β_1 will be a leading element for $A_0 = A \cap \{\alpha : \sigma(\alpha) \geq 0\}$ half space and β_3 will be a leading element for $A_1 = A \cap \{\alpha : \sigma(\alpha) < 0\}$ half space. Since $A + B = \mathbb{Z}^2$, we obtain the situation demonstrated on the next Figure 3.2. The white circular points in the

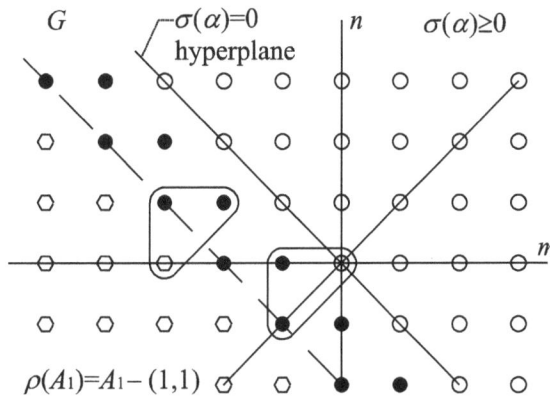

Figure 3.2: "Stirling numbers" in a different setting.

picture denote the half plain A_0 and also the image $\rho(A_0)$, since $\rho(A_0) = A_0 + \beta_1 = A_0$, the hexagon points denote the shifted half plain A_1, since $\rho(A_1) = A_1 + \beta_3 = A_1 - (1, 1)$. The black points marks the set $G = (A + B) \smallsetminus \rho(A)$ on which the initial values have to be defined. At the first stage we can calculate $y(\alpha)$ at any circular point of the $\sigma(\alpha) = 0$ hyperplane taking β_1 as leading element and a value $y(\alpha)$ at any "hexagon" point of the $\sigma(\alpha) = -3$ hyperplane taking β_3 as leading element. In the picture two examples of corresponding location of the mask are shown. From the picture it is also evident, that any function defined on G, $g : G \to \mathbb{C}$, can uniquely be extended into the solution y defined now on the whole integer plane $\mathbb{Z}^2$.

Remark 3.2.5. The proof of the 3.2.3 and also the previous examples shows that one of its assumptions, namely that all the coefficients $a(\alpha, \beta)$ should be nonzero, is unnecessarily restrictive. Evidently it is sufficient to assume that the leading coefficient is nonzero, $a(\alpha, \beta_\alpha) \neq 0$. (In fact, this has been used in Example 3.2.4.) This remark is important in the following consideration.

For linear equations the problem of existence, uniqueness and recursive construction of solution depends on set – theoretical considerations and, besides some linear operations, only one operation of division is necessary. Therefore there are no obstacles to deal with systems of linear difference equations.

It's really a simple "technical" problem to rewrite equations (3.2.2), (3.2.6) and the theorem 3.2.3 in a matrix (vector) form. For a systems of PDE's we obtain

$$\sum_{\beta \in B} \mathbb{A}(\alpha, \beta) \mathbf{y}(\alpha + \beta) = \mathbf{x}(\alpha) \tag{3.2.7}$$

where $\mathbb{A} : A \times B \to \mathbb{C}^{\mathbf{m} \times \mathbf{m}}$ are matrix coefficients, $\mathbf{x} : A \to \mathbb{C}^m$ and $\mathbf{y} : A + B \to \mathbb{C}^m$ are column vector sequences and matrix multiplication is considered (**m** means standard m – element finite set of indexes, e. g. $\mathbf{m} = \{1, 2, \dots m\}$).

Theorem 3.2.6. Let be given a system of linear difference equations (3.2.7) with non-singular matrices $\mathbb{A} : A \times B \to \mathbb{C}^{\mathbf{m} \times \mathbf{m}}$ and $A, B \subseteq \mathbb{Z}^n$, $A \neq \emptyset$, $2 \leq |B| < \infty$.

Then, there exists a set G, $G \subseteq A + B$, and a well ordering $\preceq$ of $A + B$, such that:

1. Any vector function $\mathbf{g} : G \to \mathbb{C}^m$ can be extended to a unique solution

   $\mathbf{y} \colon A + B \to \mathbb{C}^m$ of (3.2.7),

2. any value $\mathbf{y}(\alpha)$, $\alpha \in A + B$, of the solution can be recursively calculated by a finite number of arithmetic operations using only values $\mathbf{y}(\alpha')$ for $\alpha' \prec \alpha$.

As was mentioned in *Remark* 3.2.5 it is sufficient to suppose nonsingularity of leading matrices $\mathbb{A}(\alpha, \beta_\alpha)$, $\alpha \in A$. Then, inverse matrix of $\mathbb{A}(\alpha, \beta_\alpha)$ exists for every $\alpha \in A$ and the reformulation of (3.2.6) comes through evident changes.

Remark 3.2.7. If the matrix coefficients are constant and the leading matrix is singular then some special results are known, see [9]. The existence and uniqueness of solution depends on the structure of the leading matrix. Some facts are illustrated in the next example. The idea of a recursive solution has been the main tool in establishing existence and uniqueness conditions for systems of PDE's.

Example 3.2.8. Consider a system of PDE with real constant matrix coefficients $\mathbb{A} : \mathbb{N}^2 \times B \to \mathbb{R}^{\mathbf{m} \times \mathbf{m}}$, where $\mathbb{N}$ denotes natural numbers (non – negative integers), $B = \{(1, 0), (0, 1), (0, 0)\}$, $\mathbb{A}(\alpha, (1, 0)) = \mathbf{A}$, $\mathbb{A}(\alpha, (0, 1)) = \mathbf{B}$, $\mathbb{A}(\alpha, (0, 0)) = \mathbf{C}$, $\mathbf{y}, \mathbf{x} : \mathbb{N}^2 \to \mathbb{R}^m$ so

$$\mathbf{A}\mathbf{y}(i + 1, k) + \mathbf{B}\mathbf{y}(i, k + 1) + \mathbf{C}\mathbf{y}(i, k) = \mathbf{x}(i, k). \tag{3.2.8}$$

with initial values $\mathbf{g} : G \to \mathbb{R}^m$, defined on the set $G = \{(0, k) : k \in \mathbb{N}\}$.

If $\det(\mathbf{A}) \neq 0$ then there exists one and only one solution of (3.2.8) such that its restriction to G satisfies $\mathbf{y}|_G = \mathbf{g}$. The solution can be recursively calculated using the equation $\mathbf{y}(i+1,k) = \mathbf{A}^{-1}(-\mathbf{B}\mathbf{y}(i,k+1) - \mathbf{C}\mathbf{y}(i,k) + \mathbf{x}(i,k))$. In the symmetric case $(\det(\mathbf{B}) \neq 0)$ the solution of (3.2.8) results from similar considerations. For the proof see [9].

If $\det(\mathbf{A}) = 0$ and $\det(\mathbf{B}) = 0$ the results above fail. In such cases we call this system of equations a singular system. In this case the equation (3.2.8) can always be rewritten (by elementary matrix operations with rows and columns) into a canonical form

$$\tilde{\mathbf{A}}\tilde{\mathbf{y}}(i+1,k) + \tilde{\mathbf{B}}\tilde{\mathbf{y}}(i,k+1) + \tilde{\mathbf{C}}\tilde{\mathbf{y}}(i,k) = \tilde{\mathbf{I}}\tilde{\mathbf{x}}(i,k), \tag{3.2.9}$$

where $\tilde{\mathbf{A}} = \mathbf{PAQ}$, $\tilde{\mathbf{B}} = \mathbf{PBQ}$, $\tilde{\mathbf{C}} = \mathbf{PCQ}$, $\tilde{\mathbf{I}} = \mathbf{PIQ} = \mathbf{PQ}$, and $\mathbf{y} = \mathbf{Q}\tilde{\mathbf{y}}$, $\mathbf{x} = \mathbf{Q}\tilde{\mathbf{x}}$, where the matrices $\mathbf{P}$, $\mathbf{Q}$ are regular.

This relation between matrices $(\mathbf{A}, \mathbf{B}, \mathbf{C}, \ldots)$ and $(\tilde{\mathbf{A}}, \tilde{\mathbf{B}}, \tilde{\mathbf{C}}, \ldots)$, forms an equivalence, we say that the matrices (or pairs or triples of matrices) are equivalent, if there exists a pair of regular matrices $\mathbf{P}$, $\mathbf{Q}$ such that the matrix relations written above are satisfied.

Any pair of square matrices $(\mathbf{A}, \mathbf{B})$ is always equivalent to a pair $(\tilde{\mathbf{A}}, \tilde{\mathbf{B}})$ where

$$\tilde{\mathbf{A}} = \begin{bmatrix} \mathbf{I}_r & \mathbf{0} & \mathbf{0} \\ \mathbf{0} & \mathbf{N}_p & \mathbf{0} \\ \mathbf{0} & \mathbf{A}_{32} & \mathbf{0}_q \end{bmatrix}, \quad \tilde{\mathbf{B}} = \begin{bmatrix} \mathbf{B}_{11} & \mathbf{0} & \mathbf{B}_{13} \\ \mathbf{0} & \mathbf{I}_p & \mathbf{0} \\ \mathbf{0} & \mathbf{0} & \mathbf{0}_q \end{bmatrix}$$

and $\mathbf{I}_r$, $\mathbf{I}_p$ are identity square matrices of order r, p, respectively, $\mathbf{N}_p$ is a p order square nilpotent matrix, $\mathbf{0}_q$ is q order square zero matrix, other block matrices are obvious. The numbers r, p, q are uniquely determined by a pair of matrices $(\mathbf{A}, \mathbf{B})$ and $r + p + q = m$ where $p + q > 0$ in a singular case. For the proof see e.q. [10]. The existence and uniqueness of the solution of (3.2.9) now depends on various combinations of numbers r, p, q, and also on ranks of corresponding diagonal blocks of $\tilde{\mathbf{C}} = \mathbf{PCQ}$ matrix, see [9].

The significance of the Theorems 3.2.3 and 3.2.6 is in the guarantee, that the recursion based on (3.2.6) cannot fail, that the solution always exists and that it is unique for a fixed well ordering. As we can notice from examples, the sufficient conditions as formulated in Theorems 3.2.3, 3.2.6 depends on a certain interrelation among the sets A, B and G.

The "general" difference equation (3.2.1) allows for considerations similar to those in the proof of the Theorem 3.2.3, with the sets A, B known and with the resulting ordering, one of the terms in (3.2.1), say β_m, becomes the "leading element". As a corollary we obtain a rather general result: If the equation (3.2.1) can be rewritten as $y(\alpha + \beta_m) = G(\alpha, y(\alpha + \beta_1), \ldots, y(\alpha + \beta_{m-1}))$ then the conclusions on the existence and uniqueness of solution remain true for nonlinear difference equations.

It is interesting to note, that the sufficient conditions are in certain sense also necessary. It has been proved earlier in [7], that for given fixed sets A, B and G, if all linear equations (3.2.2) (satisfying some nonessential, rather technical conditions) have a unique solution, then this solution can be recursively calculated from (3.2.6).

Hexagonal grid

In many cases difference equations are understood as the result of a sampling procedure of some "continuous" quantity spreading over a plain or space. We can imagine a sampling procedure which measures this quantity in a pattern different from a rectangular grid. Regardless, sequences in PDE's are always defined on subsets of $\mathbb{Z}^n$ where the Cartesian power of $\mathbb{Z}$ is commonly understood as typically rectangular. This is not fully true, if we consider e.g. $A = \{\lambda(2,1) + \mu(1,2) : \lambda, \mu \in \mathbb{Z}\}$ and a mask $B = \{(0,0),(0,1),(1,0),(-1,-1)\}$ then the elements of $A + B$ drawing as points in the rectangular grid $\mathbb{Z}^2$ form a pattern looks like hexagonal grid as shown in following Fig. 3.4. Nevertheless, the next examples show that the "shape" of a mask or some region in $\mathbb{Z}^n$ is not meaningful unless $\mathbb{Z}^n$ is endowed with some metric.

Example 3.2.9. Let us have some quantity (function) $f : \mathbb{R}^2 \to \mathbb{R}$ measured in vertexes of an equilateral hexagonal grid of the euclidean plain $\mathbb{R}^2$, see Fig. 3.3. Further, we suppose that value of f in the central vertex X equals some weighted mean of the values at the points A, B, C surrounding the central point X and that this weights depend only on the position of the points A, B, C and X in an euclidean space.

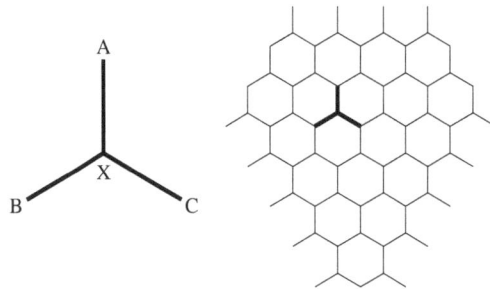

Figure 3.3: Hexagonal grid in euclidean space $\mathbb{R}^2$

For this function f we can write

$$a(X)f(X) + a(A)f(A) + a(B)f(B) + a(C)f(C) = 0. \qquad (3.2.10)$$

We want to construct a mapping of this hexagonal grid into $\mathbb{Z}^2$. Let us denote vectors $\mathbf{a} = A - X$, $\mathbf{b} = B - X$, $\mathbf{c} = C - X$ and $\mathbf{u} = \mathbf{c} - \mathbf{b}$, $\mathbf{v} = \mathbf{a} - \mathbf{b}$. The hexagonal grid can be generated by the pattern displayed on the left hand side of Fig. 3.3 by appropriate shifting along the vectors $\mathbf{u}, \mathbf{v}$. Hence all vertices's of the hexagonal grid form the set

$$H_G = \{O + \lambda\mathbf{u} + \mu\mathbf{v} + \nu\mathbf{a} : \lambda, \mu \in \mathbb{Z}, \nu \in \{0,1\}\}.$$

A fixed point O is the origin of a somewhat unusual coordinate system on this grid. Any point $X \in H_G$ uniquely determines a triple of coordinates (λ, μ, ν), where $\lambda, \mu \in \mathbb{Z}$, $\nu \in \{0,1\}$ by equation X$= O + \lambda\mathbf{u} + \mu\mathbf{v} + \nu\mathbf{a}$.

Let us define $s : H_G \to \mathbb{Z}^2$ by relations $s(X) = (m, n)$ if and only if

$$X = O + \lambda \mathbf{u} + \mu \mathbf{v} + \nu \mathbf{a} \tag{3.2.11}$$

and e. g.

$$(m, n) = \lambda(2, 1) + \mu(1, 2) + \nu(0, 1), \tag{3.2.12}$$

where the triple $(2,1)$, $(1,2)$, $(0,1)$ is one of possible choices. Now, we can convert equation (3.2.10) into a PDE. To this end we may use the injective mapping $s : H_G \to \mathbb{Z}^n$ commonly called a sampling function. This function s, as well as the dimension n, are not uniquely determined. Since any $X \in H_G$ uniquely determines (λ, μ, ν) by (3.2.11), s is a function. Since any pair of integers (m, n) from $s(H_G)$ uniquely determines (λ, μ, ν) by (3.2.12), s is an injective function having its inverse, let us denote $t = s^{-1}$. The following Fig. 3.4 illustrates the sampling s. The black points in this Figure denotes the integer pairs $(m, n) \in s(H_G)$.

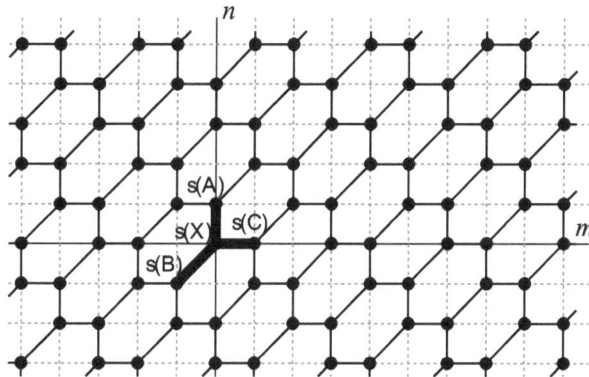

Figure 3.4: Hexagonal grid sampled with s.

Now, the conversion of (3.2.10) into a PDE is obvious. Let $X, A, B, C \in H_G$ and $s(X) = (m, n)$. Then $t(m, n) = X$ and it could be simply verified that $s(A) = s(X + \mathbf{a}) = s(X) + (0, 1)$, $s(B) = s(X + \mathbf{b}) = s(X) - (1, 1)$, $s(C) = s(X + \mathbf{c}) = s(X) + (1, 0)$. Using this relations and the inverse function t we can rewrite (3.2.10). Let us define sampled coefficients and quantities $a_s(m, n) = a \circ t(m, n)$, $f_s(m, n) = f \circ t(m, n)$, respectively, where $\circ$ denotes composition of functions. We obtain

$$\begin{aligned} a_s(m, n) \cdot f_s(m, n) + a_s(m, n + 1) \cdot f_s(m, n + 1) + \\ + a_s(m - 1, n - 1) \cdot f_s(m - 1, n - 1) + a_s(m + 1, n) \cdot f_s(m + 1, n) = 0, \end{aligned} \tag{3.2.13}$$

for any $(m, n) \in s(H_G)$. The PDE we have obtained is an equation with nonconstant coefficients, it has the mask $B_s = \{(0, 0), (0, 1), (1, 0), (-1, -1)\}$ and its solution is defined on a "hexagonal grid" $A_s = \{(m, n) : m = \lambda(2, 1) + \mu(1, 2) + \nu(0, 1)\}$ outlined in Fig. 3.4.

Of course, the sampling function s based on the relations (3.2.11), (3.2.12) is not uniquely determined. If we replace e. g. the equation (3.2.12) with the following

$$(m, n) = \lambda(2, 0) + \mu(0, 1) - \nu(1, 0), \tag{3.2.14}$$

we are constructing a different sampling function $\tilde{s}$ and obtain a different image $\tilde{s}(H_G)$ of the hexagonal grid, see Fig. 3.5.

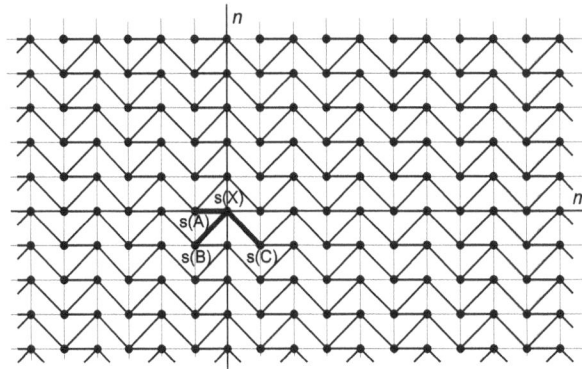

Figure 3.5: Hexagonal grid sampled with $\tilde{s}$

In this new case $\tilde{s}(H_G)$ covers the whole integer plain $\mathbb{Z}^2$, so $\tilde{s} : H_G \to \mathbb{Z}^2$ is a bijection where we have $\tilde{s}(A) = \tilde{s}(X+\mathbf{a}) = \tilde{s}(X)-(1,0)$, $\tilde{s}(B) = \tilde{s}(X+\mathbf{b}) = \tilde{s}(X)-(1,1)$, $\tilde{s}(C) = \tilde{s}(X+\mathbf{c}) = \tilde{s}(X)+(1,-1)$. Using this bijection we obtain $a_{\tilde{s}}(m,n) = a \circ \tilde{t}(m,n)$, $f_{\tilde{s}}(m,n) = f \circ \tilde{t}(m,n)$ and the PDE in the form

$$a_{\tilde{s}}(m,n) \cdot f_{\tilde{s}}(m,n) + a_{\tilde{s}}(m-1,n) \cdot f_{\tilde{s}}(m-1,n)+$$

$$\begin{aligned} +a_{\tilde{s}}(m-1,n-1) \cdot f_{\tilde{s}}(m-1,n-1)+ \\ +a_{\tilde{s}}(m+1,n-1) \cdot f_{\tilde{s}}(m+1,n-1) = 0, \end{aligned} \tag{3.2.15}$$

where $\tilde{t}$ is the inverse of $\tilde{s}$. Here we have now $A_{\tilde{s}} = \mathbb{Z}^2$ and the mask

$$B_{\tilde{s}} = \{(0,0),(-1,0),(-1,-1),(1,-1)\}.$$

The equations (3.2.13) and (3.2.15) are in an obvious sense equivalent since they originate from one equation (3.2.10). Their equivalence can be expressed explicitly using equations $f = f_s \circ s = f_{\tilde{s}} \circ \tilde{s}$ and $a = a_s \circ s = a_{\tilde{s}} \circ \tilde{s}$. We obtain $f_{\tilde{s}} = f_s \circ s \circ \tilde{t}$, $a_{\tilde{s}} = a_s \circ s \circ \tilde{t}$ where

$$s \circ \tilde{t}(m,n) = (m+n, \frac{m}{2} + 2n)$$

if m is even and

$$s \circ \tilde{t}(m,n) = (m+n+1, \frac{m+1}{2} + 2n + 1)$$

if m is odd.

Solutions of (3.2.13) or (3.2.15) along the lines of Theorem 3.2.3 yields a solution of equation (3.2.10).

Example 3.2.9 raises questions about transformations of PDE's. We can introduce the concept of equivalence of linear PDE's similarly as it is used in partial differential equations.

Boundary value problem for PDE's constitutes another important task which is solved in many applications and contexts. We will concentrate on this problem the next section.

3.2.2   Boundary value problems.

In this section we will discuss the solution of (3.2.2) under boundary condition. The boundary value problem is typically not recursively solvable in most cases. For a finite set A, the problem can be reduced to a system of linear equations, its solution, however, could be laborious. If A is infinite, some properties of the solution at infinity have to be prescribed to ensure the uniqueness of the solution. First, we concentrate on uniqueness of the solution. The problem of existence is postponed to the next subsection.

Uniqueness of the solution

In this section we again suppose for equation (3.2.2) a nonempty arbitrary set A, $\emptyset \neq A \subseteq \mathbb{Z}^n$, finite mask $B \subseteq \mathbb{Z}^n$, $2 \leq |B| < \infty$, with one restriction: we suppose that the origin $\mathbf{0} \in \mathbb{Z}^n$ is always the element of B, and the corresponding coefficient is nonzero, i.e. $\mathbf{0} \in B$, and $a(\alpha, \mathbf{0}) \neq 0$ for all $\alpha \in A$. Further, for any $\alpha \in A$ we suppose the existence of β, $\mathbf{0} \neq \beta \in B$, such that $a(\alpha, \beta) \neq 0$. Under such suppositions it is evident that $A \subseteq A + B$.

First of all we introduce a fixed set called "external boundary" $H \subseteq \mathbb{Z}^n$ on which the boundary conditions will be defined.

Definition 3.2.10. The *external boundary* corresponding to the set A and the equation (3.2.2) is the set $H = ((A + B) \smallsetminus A) \cap \{\alpha + \beta : a(\alpha, \beta) \neq 0\}$.

Remark 3.2.11. It's a simple consequence of the Definition 3.2.10 that if all the coefficients $a(\alpha, \beta)$ are nonzero (it means $a : A \times B \to \mathbb{C} \smallsetminus \{0\}$), or if we consider the equation with constant coefficients, then we have $H = (A + B) \smallsetminus A$.

We can formulate two boundary value problems.

Definition 3.2.12.

1. *Boundary value problem of the first kind.* Let $h : H \to \mathbb{C}$ be a given function. The solution of the first boundary value problem is a sequence $y : A \cup H \to \mathbb{C}$ which satisfies the equation (3.2.2) at all points $\alpha \in A$ and the boundary condition of the first kind $y(\gamma) = h(\gamma)$ for all $\gamma \in H$ and for a given function h. If, eventually, the set A is infinite, the solution has to satisfy an additional "condition at infinity"

$$\lim_{|\alpha| \to \infty} y(\alpha) = 0, \ \alpha \in A. \tag{3.2.16}$$

$\lim_{|\alpha| \to \infty} y(\alpha) = 0, \alpha \in A$ means that for any positive $\varepsilon > 0$,

$$\{\alpha : |y(\alpha)| \geq \varepsilon, \alpha \in A\}$$

is a finite set .

2. *Generalized boundary value problem.* Let two functions h, p, be given, $h, p : H \to \mathbb{C}$ and $|p(\gamma)| \leq 1$ for all $\gamma \in H$. Let $\phi : H \to A$ be a mapping of the external boundary into A. (In the simplest common cases, $\phi(\gamma)$ is a point of A neighboring

$\gamma \in H$). The solution y of the generalized boundary value problem is defined as the sequence $y : A \cup H \to \mathbb{C}$ which satisfies the equation (3.2.2) on A, boundary condition

$$y(\gamma) = p(\gamma)y\,(\phi(\gamma)) + h(\gamma) \qquad (3.2.17)$$

on H and (if A is infinite) the condition (3.2.16).

Remark 3.2.13. In the theory of partial differential equations (in the most simple cases) the boundary conditions of the second and third kind are considered in the form $\lambda y + \mu \frac{\partial y}{\partial \mathbf{v}} = f$, where $\frac{\partial}{\partial \mathbf{v}}$ denotes the directional derivative with respect to a normal vector $\mathbf{v}$ of the boundary region, oriented to the exterior of this region, and with $\lambda \geq 0$, $\mu \geq 0$ and $\lambda + \mu > 0$. The condition (3.2.17) results from a discretization of this condition, see [11].

Remark 3.2.14. In some cases, the condition (3.2.16) can be replaced by a weakened condition

$$|y(\alpha)| < M \qquad (3.2.18)$$

for all $\alpha \in A$, i.e. the sequence y is required to be bounded on A.

Remark 3.2.15. The generalized boundary value problem contains the problem of the first kind with $p(\gamma) = 0$ for all $\gamma \in H$.

For these boundary value problems the following two theorems were proved in [11].

Theorem 3.2.16. If

$$|a(\alpha, \mathbf{0})| > \sum_{\beta \in B \smallsetminus \{\mathbf{0}\}} |a(\alpha, \beta)| \qquad (3.2.19)$$

for all $\alpha \in A$, $|p(\gamma)| \leq 1$ for all $\gamma \in H$, then there exists at most one sequence y, satisfying both equation (3.2.2), the generalized boundary condition (3.2.17) and, eventually, the condition (3.2.16) at infinity of A.

To formulate the next theorem, some additional concepts have to be formulated.

Definition 3.2.17.

1. The point $\alpha \in A$ is *mask – connected* with a point $\gamma \in A \cup H$ if there is a sequence $\alpha_0, \alpha_1, \ldots, \alpha_{m-1} \in A$, $\alpha_m \in A \cup H$ such that $\alpha_0 = a$, $\alpha_m = \gamma$ and $\alpha_{k+1} - \alpha_k \in B$, $a(\alpha_k, \alpha_{k+1} - \alpha_k) \neq 0$ for $k \in \{0, \ldots, m-1\}$.

2. The nonempty set A in equation (3.2.2) will be said to have the *MC – property* with respect to the corresponding general boundary value problem if each point of A is mask – connected

   (a) either with a point of A in which the sharp inequality (3.2.19) is satisfied,

   (b) or with at least one point in each neighborhood of infinity of A,

(c) or with at least one point $\gamma \in H$ with $|p(\gamma)| < 1$ and (if $p(\gamma) \neq 0$) with its corresponding point $\phi(\gamma)$.

Theorem 3.2.18. If

$$|a(\alpha, \mathbf{0})| \geq \sum_{\beta \in B \setminus \{\mathbf{0}\}} |a(\alpha, \beta)| \tag{3.2.20}$$

for all $\alpha \in A$, $|p(\gamma)| \leq 1$ for all $\gamma \in H$, and the set A has the MC–property, then there exists at most one sequence y satisfying both the equation (3.2.2), the generalized boundary condition (3.2.17) and, eventually, the condition (3.2.16) at infinity.

Let us illustrate the notions and concepts introduced so fare with an example, which motivates the seemingly artificial definitions above.

Example 3.2.19. Let A contain only 7 points

$$A = \{(1, 1),\ (1, 2),\ (1, 3),\ (2, 1),\ (2, 2),\ (2, 3),\ (3, 2)\},$$

B contain 4 vectors $B = \{\beta_0, \beta_1, \beta_2, \beta_3\}$ where $\beta_0 = (0, 0)$, $\beta_1 = (-1, 1)$, $\beta_2 = (-1, 0)$, $\beta_3 = (-1, -1)$. Let us suppose constant coefficients of (3.2.2) $a(\alpha, \beta_0) = -3$, $a(\alpha, \beta_1) = a(\alpha, \beta_2) = a(\alpha, \beta_3) = 1$ for all $\alpha \in A$ and right hand side $x(\alpha) = 0$. Then the external boundary H contains 7 points,

$$H = (A + B) \setminus A = \{(0, 0),\ (0, 1),\ (0, 2),\ (0, 3),\ (0, 4),\ (1, 0),\ (1, 4)\},$$

(see Figure 3.6). In this Figure, the set A is represented by grayed circles, the region

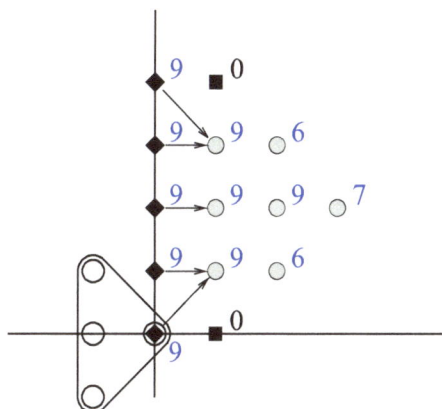

Figure 3.6: Regions in boundary value problems.

H is represented by black squares (both kinds), the picture shows also the mask in its initial (unshifted) position. In the black squares the conditions (3.2.17) are prescribed. Let us define the corresponding functions $\phi : H \to A$ and $h, p : H \to \mathbb{C}$. The function ϕ is represented by arrows in the figure, $\phi(\gamma) = \alpha$ when the arrow points from γ to α.

Consider $p(0, m) = 1$, $h(0, m) = 0$ for $m \in \{0, 1, 2, 3, 4\}$ and $p(1, 0) = p(1, 4) = 0$, $h(1, 0) = h(1, 4) = 0$. This leads to the following relations: $y(0, 0) = y(1, 1)$, $y(0, 4) =$

$y(1, 3)$, $y(0, m) = y(1, m)$ for $m \in \{1, 2, 3\}$ and $y(1, 0) = 0$, $y(1, 4) = 0$. In the last two points (represented by the squares "staying on their bases") the generalized boundary conditions are reduced to the conditions of the first kind, in the remaining points of H (squares "staying on their vertexes") boundary conditions of the second and the third kind are considered.

In the picture, the blue printed numbers show one of the solution of the problem. The solution is not unique because the equation is homogeneous and the boundary conditions also. This fact does not contradict Theorems 3.2.16, 3.2.18. The first Theorem requires the sharp inequality (3.2.19), the second requires MC–property which is not satisfied since the points $(2, 2)$, $(1, 1)$, $(1, 2)$, $(1, 3)$ are not mask–connected with any of the points $(1, 0)$ and $(1, 4)$, for which $|p| < 1$.

Fundamental solution

Many problems can be solved using the concept of the fundamental solution which characterizes several properties of the PDE's.

Definition 3.2.20. Let us suppose that the coefficients $a(\alpha, \beta)$ be defined for all $\alpha \in \mathbb{Z}^n$, so $a : \mathbb{Z}^n \times B \to \mathbb{C}$. Then, the *fundamental solution* of the equation (3.2.2) is the sequence $q : \mathbb{Z}^n \times \mathbb{Z}^n \to \mathbb{C}$ satisfying the equation

$$\sum_{\beta \in B} a(\alpha, \beta) q(\alpha + \beta, \gamma) = \delta(\alpha, \gamma), \ \alpha, \gamma \in \mathbb{Z}^n, \tag{3.2.21}$$

where $\delta : \mathbb{Z}^n \times \mathbb{Z}^n \to \{0, 1\}$ is the "unit impulse" sequence, defined as follows: $\alpha \neq \gamma \Rightarrow \delta(\alpha, \gamma) = 0$ and $\delta(\alpha, \alpha) = 1$ for any $\alpha, \gamma \in \mathbb{Z}^n$.

Without further assumptions the fundamental solution is not uniquely determined.

Theorem 3.2.21. If the coefficients in (3.2.21) satisfy (3.2.20) for all $\alpha \in \mathbb{Z}^n$ and if there exists at least one fixed nonzero element $\mathbf{0} \neq \beta \in B$ such that $a(\alpha, \beta) \neq 0$ for all $\alpha \in \mathbb{Z}^n$, then there exists at most one fundamental solution q which satisfies the condition

$$\lim_{|\alpha| \to \infty} q(\alpha, \gamma) = 0 \tag{3.2.22}$$

for each $\gamma \in \mathbb{Z}^n$.

If the coefficients $a(\alpha, \beta)$ are real, the fundamental solution q is real also.

For any fixed $\gamma \in \mathbb{Z}^n$, the sequence $\alpha \mapsto q(\alpha, \gamma)$ has its maximum at the point γ only.

Proof. The difference of two such fundamental solutions satisfy the homogeneous equation (3.2.21) on $\mathbb{Z}^n$ and the condition (3.2.22). The set B contains at least one fixed element $\mathbf{0} \neq \beta \in B$ for which $a(\alpha, \beta) \neq 0$ for all $\alpha \in \mathbb{Z}^n$. Therefore, any point of $\mathbb{Z}^n$ is mask–connected with some point in any neighborhood of infinity, so the set $\mathbb{Z}^n$ has the MC–property. Consequently, the difference of the fundamental solution has to be a zero sequence as follows from the Theorem 3.2.18.

If the coefficients a are real, the imaginary part of a complex fundamental solution satisfies the same homogeneous equation (3.2.21) and (3.2.22) and hence equals identically zero. The sequence $|q|$ cannot reach its positive maximum at any point for which the homogeneous equation is satisfied. See [11] for details.   □

In the case of the equation (3.2.21) with constant coefficients, where we can write $a(\beta) = a(\alpha, \beta)$ for all $\alpha \in \mathbb{Z}^n$ and suppose $a(\beta) \neq 0$ for all $\beta \in B$, the equation becomes "shift invariant" and the fundamental solution can be written in the form $q(\alpha, \beta) = q_s(\alpha + \beta)$, so

$$\sum_{\beta \in B} a(\beta) q_s(\alpha + \beta) = \delta(\alpha, \mathbf{0}), \ \alpha \in \mathbb{Z}^n. \tag{3.2.23}$$

Recalling the known principal properties of the fundamental solution, see [12], [13], the fundamental solution exists in many cases and can be written in the integral form

$$q_s(\alpha) = \frac{1}{(2\pi)^n} \int_{K^n} \frac{exp(i\alpha \cdot \kappa)}{D(\kappa)} d\kappa, \tag{3.2.24}$$

where $K^n = \langle -\pi, \pi \rangle^n$ is a n-dimensional cube, $\kappa \in K^n$ is a n-dimensional real vector, $\alpha \cdot \kappa = \alpha^1 \kappa^1 + \cdots + \alpha^n \kappa^n$ means the dot product and $D : K^n \to \mathbb{C}$ is a function determined by

$$D(\kappa) = \sum_{\beta \in B} a(\beta) e^{i\beta \cdot \kappa}. \tag{3.2.25}$$

If

$$|a(\mathbf{0})| > \sum_{\beta \in B \smallsetminus \{\mathbf{0}\}} |a(\beta)| \tag{3.2.26}$$

then $D(\kappa) \neq 0$ on K^n, therefore, the integral (3.2.24) converges absolutely, satisfies (3.2.23) and the condition at infinity (3.2.22),

$$\lim_{|\alpha| \to \infty} q_s(\alpha) = 0. \tag{3.2.27}$$

When

$$|a(\mathbf{0})| = \sum_{\beta \in B \smallsetminus \{\mathbf{0}\}} |a(\beta)|, \tag{3.2.28}$$

the construction and properties of the fundamental solution are more complicated. If still $D(\kappa) \neq 0$ on K^n then the solution can be written also in the form (3.2.24) and (3.2.27) is satisfied. If the function (3.2.25) has some zero points the form of the integral has to be modified. Some modifications and special results are published in [13].

Existence of the solution

In this section we will use the Theorems 3.2.16, 3.2.18 to show the existence and to construct the solution of (3.2.2) under boundary conditions (3.2.17).

If A is finite, the difference equation with the boundary conditions represents a finite system of linear equations with a diagonally semi–dominant square matrix. Under the conditions of Theorem 3.2.16 or Theorem 3.2.18, the corresponding homogeneous problem has only a zero solution. Therefore, the non–homogeneous problem has one and only one solution.

If A is not finite, we will formulate general existence theorems based on properties of the fundamental solution. For details see [11].

Definition 3.2.22. Let us suppose that the coefficients $a(\alpha, \beta)$ are defined for all $\alpha \in \mathbb{Z}^n$, so $a : \mathbb{Z}^n \times B \to \mathbb{C}$. For each boundary value problem on $A \subseteq \mathbb{Z}^n$, $A \neq \mathbb{Z}^n$ we define an *auxiliary homogeneous boundary value problem* as a problem of solution of (3.2.2) on the set $A_0 = \mathbb{Z}^n \smallsetminus (A \cup H)$ with homogeneous boundary conditions of the first kind on a set H_0, so $p(\gamma) = 0$, $h(\gamma) = 0$ for all $\gamma \in H_0$, where

$$H_0 = ((A_0 + B) \smallsetminus A_0) \cap \{\alpha + \beta : a(\alpha, \beta) \neq 0\}.$$

If (3.2.2) has constant coefficients, $a(\alpha, \beta) = a(\beta)$, we have $H_0 = (A_0 + B) \smallsetminus A_0$.

Theorem 3.2.23. (Existence of the solution) Let us suppose that one of the two sets A and $\mathbb{Z}^n \smallsetminus A$ be finite. Let the conditions for the uniqueness of the solution of (3.2.2) on A under generalized boundary conditions (3.2.17) on H and *(if A is infinite)* under the condition (3.2.16) be fulfilled *(suppositions of the Theorems 3.2.19 or 3.2.20)*.

If the set $A_0 = \mathbb{Z}^n \smallsetminus (A \cup H)$ is not empty, suppose that the conditions for uniqueness of the solution of the corresponding auxiliary homogeneous boundary value problem are satisfied.

Suppose that x is a sequence having finite number of nonzero values and that there exist a fundamental solution q of (3.2.2) on $\mathbb{Z}^n$ with the property $\lim_{|\alpha| \to \infty} q(\alpha, \gamma) = 0$ for each $\gamma \in H \cup H_0$.

Under these conditions, there exists one and only one solution of (3.2.2) on A under boundary conditions (3.2.17) and, eventually *(if A is infinite)*, condition (3.2.16).

Proof, examples as well as other rather complicated statements can be found in [11].

3.3 Growth estimates for PDE's.

Boundedness and growth estimates for solutions is an important issue for difference equations. For equations with constant coefficients a vast amount of papers deals with conditions for the solution of (3.2.2) to be bounded, mainly for the case $A = \mathbb{N}^n$. These conditions are often not easy to verify. For this special case and linear PDE's with variable coefficients a method similar to our results has been published in [14]. The case $A = \mathbb{Z}^n$ has been extensively dealt with in [15] and in papers cited therein.

Subsequent considerations deal with a general case under conditions which are used several times in the next theorems.They are summarized here for convenience.

Assumptions.

1. $A, B \subseteq \mathbb{Z}^n$, $A \neq \emptyset$, $2 \leq |B| < \infty$.

2. For any $\alpha \in A$ there is a leading coefficient β_α such that $a(\alpha, \beta_\alpha) \neq 0$.

3. The relation $\alpha \mapsto \beta_\alpha$ is an injective mapping $\rho : A \to A + B$, $\rho(\alpha) = \alpha + \beta_\alpha$.

4. The initial value set G is defined by $G = (A + B) \smallsetminus \rho(A)$.

5. There is an order $\preccurlyeq$ such that $(A, \preccurlyeq)$ is well ordered set and for any $\alpha \in A$ it is true

$$\alpha + B_\alpha \subseteq \Gamma_\alpha \cup G, \tag{3.3.1}$$

where $B_\alpha = B \smallsetminus \{\beta_\alpha\}$, $(\leftarrow, \alpha) = \{\alpha' \in A : \alpha' \prec \alpha\}$ and $\Gamma_\alpha = \rho((\leftarrow, \alpha))$.

6. A solution $y : A + B \to \mathbb{C}$ of (3.2.2) can be recursively calculated by (3.2.6).

In fact, Theorem 3.2.3 shows that if the Assumptions 1 and 2 are satisfied then other items are a consequence of them. The Assumptions 1 and 2 guarantee the existence and uniqueness of the solution of (3.2.2) recursively calculated by (3.2.6) and also the existence of sets, mappings and relations described in items 3 – 5.

It is useful to introduce a *supreme norm* for sequences, as e. g. $x : A \to \mathbb{C}$, defined by

$$\|x\|_A = \sup_{\alpha \in A} |x(\alpha)|.$$

The norm depends on the set where the supremum is considered. It is obvious that the following statements are simple consequences of the definition: $|x(\alpha)| \leq \|x\|_S$ for any $\alpha \in S$, $S \subseteq T \Rightarrow \|x\|_S \leq \|x\|_T$ and $\|x\|_{S \cup T} = max\{\|x\|_S, \|x\|_T\}$. This is valid also for any l_p norm $\|x\|_A = \left(\sum_{\alpha \in A} |x(\alpha)|^p \right)^{1/p}$ except for the last statement, where we have only inequality $\|x\|_{S \cup T} \geq max\{\|x\|_S, \|x\|_T\}$.

The following theorems are slightly more general compared to those in [16] and therefore they are presented here with complete proofs.

Theorem 3.3.1. *(The maximum principle).* Consider a homogeneous equation (3.2.2) $(x \equiv 0)$ under the Assumptions 1 – 6.

If the coefficients of (3.2.2) hold the relation

$$|a(\alpha, \beta_\alpha)| \geq \sum_{\beta \in B_\alpha} |a(\alpha, \beta)| \quad \text{for all} \quad \alpha \in A \tag{3.3.2}$$

then any solution $y : A + B \to \mathbb{C}$ of homogeneous equation (3.2.2) satisfies the inequality

$$|y(\kappa)| \leq \sup_{\gamma \in G} |y(\gamma)| \quad \text{for all} \quad \kappa \in A + B.$$

Proof. Equation (3.2.2) with the relation (3.3.2) can evidently be considered in the reduced form

$$y(\alpha + \beta_\alpha) = \sum_{\beta \in B_\alpha} b(\alpha, \beta) y(\alpha + \beta), \qquad (3.3.3)$$

where $b(\alpha, \beta) = -a(\alpha, \beta)/a(\alpha, \beta_\alpha)$ and $\sum_{\beta \in B_\alpha} |b(\alpha, \beta)| \le 1$.

From (3.3.3) we obtain

$$|y(\alpha + \beta_\alpha)| \le \sum_{\beta \in B_\alpha} |b(\alpha, \beta)| \, |y(\alpha + \beta)| \le \|y\|_{\alpha + B_\alpha} \sum_{\beta \in B_\alpha} |b(\alpha, \beta)| \le \|y\|_{\Gamma_\alpha \cup G},$$

where we denote $\Gamma_\alpha = \rho((\leftarrow, \alpha))$ for simplicity, so we have

$$|y(\alpha + \beta_\alpha)| \le \|y\|_{\Gamma_\alpha \cup G}, \quad \text{for any} \quad \alpha \in A. \qquad (3.3.4)$$

First we show by (transfinite) induction that

$$|y(\alpha + \beta_\alpha)| \le \|y\|_G, \quad \text{for any} \quad \alpha \in A. \qquad (3.3.5)$$

The statement (3.3.5) is evidently valid for the the least element α_0 of A, $\alpha_0 \in A$. It is due to (3.3.4) and the fact that $\Gamma_{\alpha_0} = \emptyset$.

Let now $\alpha_0 \prec \alpha$ and $|y(\alpha' + \beta_{\alpha'})| \le \|y\|_G$, for any $\alpha' \prec \alpha$. Then, we have $\sup_{\alpha' \prec \alpha} |y(\alpha' + \beta_{\alpha'})| \le \|y\|_G$, so $\|y\|_{\Gamma_\alpha} \le \|y\|_G$, and $\|y\|_{\Gamma_\alpha \cup G} = \|y\|_G$. Finally, taking (3.3.4) into account, we obtain $|y(\alpha + \beta_\alpha)| \le \|y\|_G$. So, (3.3.5) has been proved.

The relation (3.3.5) can be written in the form $\|y\|_{\rho(A)} \le \|y\|_G$ from which we have $\|y\|_{\rho(A) \cup G} = \|y\|_G$ and $\|y\|_{A+B} = \|y\|_G$ by Assumption 4. $\qquad \square$

Remark 3.3.2. The relation (3.3.2) imposed on the coefficients of (3.2.2) is rather restrictive. This restriction can be weakened by a simple substitution. If for a chosen sequence $w : A + B \to \mathbb{C}$ a new sequences $z : A + B \to \mathbb{C}$ is defined by equality $y(\gamma) = z(\gamma) w(\gamma)$, then the sequence $z : A + B \to \mathbb{C}$ satisfies equation (3.2.2) with new coefficients $\tilde{a}(\alpha, \beta) = a(\alpha, \beta) w(\alpha + \beta)$.

If we chose e. g. $w(\gamma) = \lambda^\gamma$, we obtain as a direct corollary of the Theorem 3.3.1 the following:

Theorem 3.3.3. Consider a homogeneous equation (3.2.2) ($x \equiv 0$) under the Assumptions $1 - 6$.

If the coefficients of (3.2.2) satisfy the relation

$$|a(\alpha, \beta_\alpha)| \lambda^{\beta_\alpha} \ge \sum_{\beta \in B_\alpha} |a(\alpha, \beta)| \lambda^\beta \quad \text{for all} \quad \alpha \in A \qquad (3.3.6)$$

then any solution $y : A + B \to \mathbb{C}$ of a homogeneous equation (3.2.2) satisfies the inequality

$$|y(\kappa)| \le \lambda^\kappa \sup_{\gamma \in G} |y(\gamma) \lambda^{-\gamma}| \quad \text{for all} \quad \kappa \in A + B,$$

where we denote $\lambda^\kappa = \lambda_1^{\kappa_1} \cdots \lambda_n^{\kappa_n}$, $\lambda \in \mathbb{R}^n$.

Proof. Any solution $y : A + B \to \mathbb{C}$ of the homogeneous equation (3.2.2) can be written in the form $y(\gamma) = z(\gamma)\lambda^{\gamma}$. Then we obtain $\sum_{\beta \in B} a(\alpha, \beta)z(\alpha + \beta)\lambda^{\alpha + \beta} = 0$. The sequence $z : A + B \to \mathbb{C}$ solves the equation $\sum_{\beta \in B} \tilde{a}(\alpha, \beta)z(\alpha + \beta) = 0$ with the coefficients $\tilde{a}(\alpha, \beta) = a(\alpha, \beta)\lambda^{\beta}$. Because of (3.3.6) the coefficients $\tilde{a}$ satisfies (3.3.2). For the solution $z : A + B \to \mathbb{C}$ the Theorem 3.3.1 gives immediately $\|z\|_{A+B} = \|z\|_G$, and we obtain the result. $\qquad\square$

The growth of a solution can be estimated also by comparison of two solutions of similar PDE's.

Theorem 3.3.4. Consider two sequences $u : A + B \to \mathbb{R}$, $v : A + B \to \mathbb{R}$ satisfying the homogeneous equations

$$u(\alpha + \beta_\alpha) = \sum_{\beta \in B_\alpha} p(\alpha, \beta)u(\alpha + \beta), \qquad (3.3.7)$$

$$v(\alpha + \beta_\alpha) = \sum_{\beta \in B_\alpha} q(\alpha, \beta)v(\alpha + \beta), \qquad (3.3.8)$$

under the Assumptions 1 – 6.

If the coefficients $p, q : A \times B \to \mathbb{R}$ are real, $0 \le p(\alpha, \beta) \le q(\alpha, \beta)$ for all $\alpha \in A$, $\beta \in B_\alpha$ and $0 \le u(\gamma) \le v(\gamma)$ for all $\gamma \in G$, then

$$0 \le u(\gamma) \le v(\gamma), \text{ for } \forall all \ \gamma \in A + B.$$

Proof. By contradiction, let there exist $\gamma_0 \in A + B$ such that either $u(\gamma_0) < 0$ or $u(\gamma_0) > v(\gamma_0)$. Since $0 \le u(\gamma) \le v(\gamma)$ for all $\gamma \in G$ then $\gamma_0 \in (A + B) \smallsetminus G = \rho(A)$ and there is $\alpha_0 \in A$ such that either $u(\alpha_0 + \beta_{\alpha_0}) < 0$ or $u(\alpha_0 + \beta_{\alpha_0}) > v(\alpha_0 + \beta_{\alpha_0})$.

Case 1. Let $\alpha_0 \in A$ be the least element for which $u(\alpha_0 + \beta_{\alpha_0}) < 0$. Then $\alpha \prec \alpha_0 \Rightarrow u(\alpha + \beta_\alpha) \ge 0$, which is equivalent to the statement $\gamma \in \rho((\leftarrow, \alpha_0)) \Rightarrow u(\gamma) \ge 0$. Since for all $\gamma \in G$ we have $u(\gamma) \ge 0$ also, we can write $\gamma \in \rho((\leftarrow, \alpha_0)) \cup G \Rightarrow u(\gamma) \ge 0$. This result together with (3.3.1) results in negativeness of r. h. s. of (3.3.7), so we obtain $u(\alpha_0 + \beta_{\alpha_0}) \ge 0$ which completes the contradiction. $0 \le u(\gamma)$ for all $\gamma \in A + B$ has been proved.

Case 2. Let $\alpha_0 \in A$ be the least element for which $u(\alpha_0 + \beta_{\alpha_0}) > v(\alpha_0 + \beta_{\alpha_0})$. Then $\alpha \prec \alpha_0 \Rightarrow u(\alpha + \beta_\alpha) \le v(\alpha + \beta_\alpha)$ which is equivalent to the statement $\gamma \in \rho((\leftarrow, \alpha_0)) \Rightarrow u(\gamma) \le v(\gamma)$. Since for all $\gamma \in G$ we have $u(\gamma) \le v(\gamma)$ also, we can write $\gamma \in \rho((\leftarrow, \alpha_0)) \cup G \Rightarrow u(\gamma) \le v(\gamma)$. This result together with (3.3.1) and the previous case gives inequalities $0 \le p(\alpha_0, \beta)u(\alpha_0 + \beta) \le q(\alpha_0, \beta)v(\alpha_0 + \beta)$ for all $\beta \in B_{\alpha_0}$. Finally, recursive calculations (3.3.7), (3.3.8) give inequality $u(\alpha_0 + \beta_{\alpha_0}) \le v(\alpha_0 + \beta_{\alpha_0})$ which completes the contradiction. Hence, $u(\gamma) \le v(\gamma)$ for all $\gamma \in A + B$ has been proved. $\qquad\square$

Under stronger conditions, we can estimate the growth of solutions of nonhomogeneous equation (3.2.2).

Theorem 3.3.5. Consider the equation (3.2.2) under the Assumptions 1 – 6. Define the numbers

$$K = \sup_{\alpha \in A} \left| \frac{1}{a(\alpha, \beta_\alpha)} \right|, \quad L = \sup_{\alpha \in A} \sum_{\beta \in B_\alpha} \left| \frac{a(\alpha, \beta)}{a(\alpha, \beta_\alpha)} \right|.$$

If $L < 1$ then the following relations are satisfied:

$$\|y\|_{\rho(A)} \leq \frac{K}{1-L} \|x\|_A + \frac{L}{1-L} \|y\|_G$$

and

$$\|y\|_{A+B} \leq \max\{\|y\|_{\rho(A)}, \|y\|_G\}.$$

Proof. From the equation (3.2.6) we simply obtain by the triangular inequality and the properties of supremum the relation $|y(\alpha + \beta_\alpha)| \leq K \|x\|_A + L \|y\|_{\alpha+B_\alpha}$ for all $\alpha \in A$. Since $\alpha + B_\alpha \subseteq A + B = \rho(A) \cup G$, we have $|y(\alpha + \beta_\alpha)| \leq K \|x\|_A + L \|y\|_{\rho(A) \cup G}$ for all $\alpha \in A$ and $\|y\|_{\rho(A) \cup G} = \max\{\|y\|_{\rho(A)}, \|y\|_G\} \leq \|y\|_{\rho(A)} + \|y\|_G$. Then $\|y\|_{\rho(A)} \leq K \|x\|_A + L(\|y\|_{\rho(A)} + \|y\|_G)$ and the rest is obvious. □

We will show the application of these estimates in the next section. The use of the maximum principle can also be demonstrated e. g. by obtaining estimates of growth for Stirling numbers or other quantities defined by partial difference equations.

Example 3.3.6. (Growth estimates for Stirling numbers)

Consider the Example 3.2.2 where the equations for a recursive calculation of the Stirling numbers have been introduced. Both equations (3.2.4) and (3.2.5) have the constant (independent on $\alpha \in A$) leading element $\beta_1 = (0,0)$, the equation (3.2.4) has the coefficients $a(\alpha, \beta_1) = 1$ (leading coefficient), $a(\alpha, \beta_2) = \alpha^1 - 1$ and $a(\alpha, \beta_3) = -1$. This coefficients do not satisfy the condition (3.3.2) of the Theorem 3.3.1 since $1 \geq |\alpha^1 - 1| + 1$ is not valid for all $\alpha \in A$. The equation (3.2.5) has similarly $a(\alpha, \beta_1) = 1$ (leading coefficient), $a(\alpha, \beta_2) = -\alpha^2$ and $a(\alpha, \beta_3) = -1$. For this equation the relation (3.3.2) is not satisfied also, we have $1 \not\geq |\alpha^2| + 1$ for some $\alpha \in A$.

As it was suggested in the *Remark* 3.3.2, we can introduce a new unknown sequence $z : A + B \to \mathbb{C}$ by the equation $y(\gamma) = z(\gamma)w(\gamma)$.

Case 1. (Stirling numbers of the first kind). Let us adopt the notation from the Example 3.2.2 where $\alpha = (m, n)$ and introduce a new sequence by equation

$$y(m, n) = m! \cdot z(m, n).$$

We obtain a new equation with the same initial set G and the same initial values for the sequence z. For all $(m, n) \in A$ we have:

$$z(m, n) = \frac{1-m}{m} z(m-1, n) + \frac{1}{m} z(m-1, n-1) \text{ and } z_{|G} = y_{|G}.$$

The new coefficients now satisfy the inequality of the Theorem 3.3.1, $1 \geq \left| \frac{1-m}{m} \right| + \left| \frac{1}{m} \right| = 1$, so we have estimation $|z(m,n)| \leq \sup_{(m,n) \in G} |z(m,n)| = 1$ for all $(m, n) \in A + B$. Back substitution gives

$$|y(m, n)| \leq m!$$

for all $(m, n) \in A \Leftrightarrow (m, n) \in \mathbb{Z}^2$, $0 \leq m$, $0 \leq n \leq m$.

Case 2. (Stirling numbers of the second kind). An appropriate substitution can be found by adjusting a real constant $\lambda > 0$ and ν in the substitution formula

$$y(m, n) = (\lambda(n + 1))^{m-n+\nu n} z(m, n)$$

such that (3.3.2) will be satisfied. After the substitution into (3.2.5) we obtain the equation

$$z(m, n) = \tilde{b}_2(m, n)z(m - 1, n) + \tilde{b}_3(m, n)z(m - 1, n - 1)$$

and also $z_{|G} = y_{|G}$, where $\tilde{b}_2(m, n) = \frac{n}{\lambda(n+1)}$, $\tilde{b}_3(m, n) = \frac{1}{\left[\lambda n \left(1 + \frac{1}{n}\right)^n\right]^\nu}\left(\frac{n}{n+1}\right)^{m-n}$. It is clear that $1 \geq \left|\tilde{b}_2(m, n)\right| + \left|\tilde{b}_3(m, n)\right|$ for all $(m, n) \in A$ if and only if $1 \geq \frac{n}{\lambda(n+1)} + \frac{1}{\left[\lambda n \left(1 + \frac{1}{n}\right)^n\right]^\nu}$ for all $n \in \mathbb{N}$, $n \geq 1$. When we chose $\nu = 1/2$ and because of the first term we need $\lambda \geq 1$, the sequence in a question is decreasing and for its greatest value we have $1 \geq \frac{1}{\lambda \cdot 2} + \frac{1}{\sqrt{\lambda \cdot 2}}$ from which for the smallest λ we find $\lambda = \frac{1}{4}(3 + \sqrt{5}) \approx 1.31$. So, we can conclude

$$y(m, n) \leq \left(\frac{3 + \sqrt{5}}{4}(n + 1)\right)^{m - \frac{1}{2}n}$$

for all $(m, n) \in A \Leftrightarrow (m, n) \in \mathbb{Z}^2$, $0 \leq m$, $0 \leq n \leq m$.

3.4   Application in multidimensional discrete systems theory.

A system, defined as a subset $\mathcal{S}$ of the Cartesian product $\mathcal{A} \times \mathcal{B}$ is commonly called a discrete system if $\mathcal{A}$, $\mathcal{B}$ are sets of functions (signals) defined on countable sets. Elements of $\mathcal{A}$ and $\mathcal{B}$ are called input and output of the system, respectively. The fact that $\mathcal{S} \subseteq \mathcal{A} \times \mathcal{B}$ we will write also as $\mathcal{S} : \mathcal{A} \to \mathcal{B}$, or $\mathcal{A} \xrightarrow{\mathcal{S}} \mathcal{B}$.

The best known example of discrete systems is that when this countable set is the set of non–negative integers, the input and output are sequences and the relation $\mathcal{S}$ is defined by a linear or nonlinear equation in finite differences, with its solution defined as the output of the system. Decades of efforts in developing the theory of linear systems of this type revealed both new methods (such as the Z–transform, stability investigations, frequency analysis and Fourier methods, etc.) and new concepts (such as state space models, their causality, time–invariance controllability, observability etc.). These concepts reflect properties of mathematical models of real–word systems, which are indispensable in applications. Results of the theory of these linear discrete systems, together with the classical theory of ordinary differential equations contributed to a general (abstract) theory of linear systems. Here a firm basis of fundamental concepts of a systems theory has been given together with their motivation, corollaries and further specifications. However, it turned out that a "direct" generalization of the (one – dimensional) systems theory concepts to the n – dimensional case has some inherent limitations.

For one – dimensional linear discrete systems one of the most important concepts is their time – invariance. On the set of sequences $\mathcal{A} = \mathbb{C}^{\mathbb{Z}}$ define an operator T^{σ} for any $\sigma \in \mathbb{Z}$, mapping the set $\mathcal{A}$ into itself as follows $(T^{\sigma}x)(\alpha) = x(\alpha - \sigma)$ for all $\alpha \in \mathbb{Z}$. A linear system $\mathcal{S} \subseteq \mathcal{A} \times \mathcal{B}$ is called time – invariant if for any $(x, y) \in \mathcal{S}$ there is $(T^{\sigma}x, T^{\sigma}y) \in \mathcal{S}$. A linear system $\mathcal{S} \subseteq \mathbb{C}^{\mathbb{Z}} \times \mathbb{C}^{\mathbb{Z}}$ defined by a difference equation is time invariant if and only if all coefficients of the difference equation are constant. (We will be more precise lately.) To characterize a linear system itself, irrespectively of its inputs, commonly the concept of impulse response or transfer function is used. If the system is time invariant then this impulse response h allows to express the output y by an operation $*$ called convolution as follows $y = h * x$, where x is the input and h is independent of x.

It is well known e. g. from the theory of $\mathbb{Z}$ – transform that systems defined on the set of "one – sided" sequences, i. e. with $\mathcal{A} \subseteq \mathbb{C}^{A}$, $A = \{n \in \mathbb{Z} : n \geq 0\}$, two types of the above operator T^{σ} have to be distinguished – the forward and backward shift. Their different meaning stems from the assumption that any signal $x \in \mathcal{A}$ shifted forward is supposed to be zero for all $\alpha \notin A + \sigma$. As a consequence also the definition of convolution, which is an important binary operation on the set of signals, seems to be different. While for sequences defined on the set $\mathbb{Z}$ the convolution of two signals x, y can be defined as the signal $z(n) = \sum_{k=-\infty}^{+\infty} x(k) \, y(n - k)$, with sequences, which are assumed to be zero for all negative values of their independent variable commonly the following definition of convolution is adopted $z(n) = \sum_{k=0}^{n} x(k) \, y(n - k)$.

In systems theory various types of functional transforms are often applied. It is worth noting that all these functional transforms have one property in common: they are of the so called convolutional type, i. e. all they transform the convolution of originals into product of images. This is one of the reasons why it is important to carefully generalize the above concept of time invariance. Note also that the term time – invariance implies a certain physical interpretation of the independent variable.

We want to generalize the basic results of the one dimensional discrete systems theory to n – dimensional discrete systems. These systems will be here considered as described by partial difference equations of the form

$$\sum_{\beta \in B} a_{\beta} y(\alpha + \beta) = x(\alpha), \quad \alpha \in A, \tag{3.4.1}$$

it means $(x, y) \in \mathcal{S} \Leftrightarrow y$ solves (3.4.1) with right hand side x, so $\mathcal{S} \subseteq \mathbb{C}^{A} \times \mathbb{C}^{A+B}$. The results on recursive computation of solution of some functional equations may point toward a way of generalization of the time concept and the concept of time – invariance. We want also to show how is time invariance and other results on PDE's connected to stability, which is one of the principal topics in system theory.

To this end we will introduce two types of operators and systems by the following definitions:

Definition 3.4.1. Let $A \subseteq \mathbb{Z}^{n}$.

For any $\sigma \in \mathbb{Z}^n$ let the operator S^σ map $\mathbb{C}^A$ into $\mathbb{C}^{A+\sigma}$ as follows: for $x \in \mathbb{C}^A$, $S^\sigma x \in \mathbb{C}^{A+\sigma}$ is defined by $(S^\sigma x)(\alpha + \sigma) = x(\alpha)$ for all $\alpha \in A$.

For any $\sigma \in \mathbb{Z}^n$ let the operator T^σ map $\mathbb{C}^A$ into $\mathbb{C}^A$ as follows, for all $\alpha \in A$

$$(T^\sigma x)(\alpha) = \begin{cases} x(\alpha - \sigma) & \text{for all } \alpha - \sigma \in A, \\ 0 & \text{for all } \alpha - \sigma \notin A. \end{cases}$$

Henceforth, S^σ is called the *shift* operator and T^σ is called the *translation* operator . If $A = A + \sigma$, (e.g. if $A = \mathbb{Z}^n$) then $T^\sigma = S^\sigma$. For the translation operator some "degenerate" cases might occur: if $A \cap (A + \sigma) = \emptyset$ then $(T^\sigma x)(\alpha) = 0$ for all $\alpha \in A$ since no $\alpha \in A$ with $\alpha - \sigma \in A$ exists. This type of translations will be excluded in the sequel.

Definition 3.4.2.

- A system $\mathcal{S} \subseteq \mathcal{A} \times \mathcal{B}$ is called $\sigma-shift-invariant$ and $\sigma-translation-invariant$ if $(x,y) \in \mathcal{S}$ implies $(S^\sigma x, S^\sigma y) \in \mathcal{S}$ and $(T^\sigma x, T^\sigma y) \in \mathcal{S}$, respectively.
  These concepts of invariant systems can be equivalently introduced by the following commutative diagrams,

$$
\begin{array}{ccc}
\mathbb{C}^A & \xrightarrow{S^\sigma} & \mathbb{C}^{A+\sigma} \\
\downarrow \mathcal{S} & & \downarrow \mathcal{S} \\
\mathbb{C}^{A+B} & \xrightarrow{S^\sigma} & \mathbb{C}^{A+B+\sigma}
\end{array}
\qquad
\begin{array}{ccc}
\mathbb{C}^A & \xrightarrow{T^\sigma} & \mathbb{C}^A \\
\downarrow \mathcal{S} & & \downarrow \mathcal{S} \\
\mathbb{C}^{A+B} & \xrightarrow{T^\sigma} & \mathbb{C}^{A+B}
\end{array}
$$

  where comutativeness of the diagrams means the equality of compositions $\mathcal{S} \circ S^\sigma = S^\sigma \circ \mathcal{S}$, $\mathcal{S} \circ T^\sigma = T^\sigma \circ \mathcal{S}$, for $\sigma-$shift–invariant and $\sigma-$translation–invariant system, respectively.

- A system $\mathcal{S} \subseteq \mathcal{A} \times \mathcal{B}$ is called translation–invariant if it is $\sigma-$translation invariant for all σ such that $A + \sigma \subseteq A$.

For shift invariant systems we assume that initial conditions $g \in \mathbb{C}^G$ for the shifted system become $S^\sigma g \in \mathbb{C}^{G+\sigma}$, while for translation–invariant systems we assume that $A \cap (A + \sigma)$ is nonempty. The last assumption is satisfied for every $\sigma \in A$ if A is a semigroup, then $A + \sigma \subseteq A$.

The following Figure 3.7 illustrates the translation invariant system.

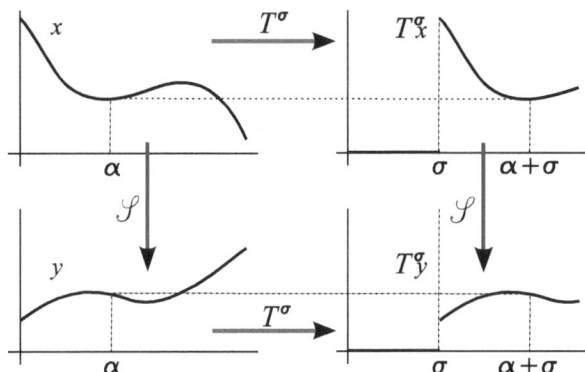

Figure 3.7: Translation invariant system.

For a linear system $\mathcal{S} \subseteq \mathbb{C}^A \times \mathbb{C}^A$ defined on a set $A \subseteq Z^n$ by a difference equation (3.2.2) to be shift – or translation – invariant it is evidently necessary that the coefficients of the equation are constant. This condition is not sufficient as shown by the following example.

Example 3.4.3. Let $A = \{(\alpha^1, \alpha^2) : \alpha^1 \geq 0\} \subseteq \mathbb{Z}^2$. Consider the equation

$$\sum_{\beta \in B} y(\alpha + \beta) = \delta(\alpha), \quad y(\gamma) = 0 \text{ for all } \gamma \in G,$$

with $B = (0,0), (0,1), (1,0), (1,-1)$ and with the initial set G, $G = \{(\alpha^1, \alpha^2) : \alpha^1\alpha^2 = 0, \alpha^1 \geq 0\}$. It can be easily verified that the output of the system defined as a solution of this equation remains unchanged when its input is changed to $T^\sigma\delta$ with $\sigma = (-1, 1)$ or $\sigma = (1, 0)$, hence the system is not translation invariant although the difference equation has constant coefficients.

Detailed analysis of the translation operator (see e.g. [17]) yields the following sufficient conditions.

Theorem 3.4.4. If in equation (3.4.1) the "leading element" β_0 is constant for all $\alpha \in A$ and

$$A + \beta_0 \subseteq \bigcap_{\beta \in B} (A + \beta)$$

then the solution of the equation is translation – invariant.

In what follows we want to find conditions enabling to generalize the concept of impulse response and convolution for n – D systems defined by equation (3.4.1).

An important property of (one – dimensional) discrete convolution $h * x$ defined by $y(n) = \sum_{\gamma \in \mathbb{N}} h(\gamma) x(n - \gamma)$ is that for any n the sum consists of a finite number of nonzero summands when both h and x are assumed to be zero for all $n < 0$. (Note that $\mathbb{N}$ is a semigroup.) To preserve this property for sequences defined on subsets of $\mathbb{Z}^n$ we will define the following concept: A subset $M \subseteq \mathbb{Z}^n$ is called an F – subset of the semigroup $\mathbb{Z}^n$ if to any of its element γ there exist at most a finite number of pairs

(α, β) such that $\alpha + \beta = \gamma$. Note that some F – subsets of $\mathbb{Z}^n$ can be semigroups . It can be proved by contradiction that for an F – semigroup $A \subseteq \mathbb{Z}^n$ and a finite set $B \subseteq \mathbb{Z}^n$ the set $A + B$ is an F – subset of $\mathbb{Z}^n$. Now the definition of convolution of sequences defined on subsets of $\mathbb{Z}^n$ can be formulated.

Definition 3.4.5. Let G be an F – semigroup of $\mathbb{Z}^n$ and let $H \subseteq \mathbb{Z}^n$ be a finite set. Consider the sequences $g : G \to \mathbb{C}$, $h : G + H \to \mathbb{C}$ then the sequence $f : G + H \to \mathbb{C}$ defined by

$$f(\alpha) = \sum_{\gamma \in C_\alpha} g(\gamma) h(\alpha - \gamma),$$

where $C_\alpha = \{\gamma : \gamma \in G, \ \alpha - \gamma \in G + H\}$ is called the convolution of g and h on the set G, i.e. $f = g * h$.

Although some common properties of convolution have to be here handled with care if $G \neq G + H$ the basic result of systems theory on impulse response can be proved. It is formulated by the following

Theorem 3.4.6. Let equation (3.4.1) be given with "zero initial conditions", i.e. with $y(\alpha) = 0$ for all $\alpha \in G$. If A is an F – semigroup, well ordered as in Theorem 3.2.3 and the sequence $h : A + B \to \mathbb{C}$ is such that

$$\sum_{\beta \in B} a_\beta h(\alpha + \beta) = \delta \quad \text{for all } \alpha \in A, \tag{3.4.2}$$

$$h(\alpha) = 0 \quad \text{for all } \alpha \in G$$

then the following statements are equivalent:

- The function $y = h * x$ is a (unique) solution of equation (3.4.1) for any $x : A \to \mathbb{C}$.

- Solutions y of equation (3.4.1) are translation invariant.

Here
$$\delta(\alpha) = \begin{cases} 1 & \text{for the least element } \alpha \text{of } A \\ 0 & \text{otherwise} \end{cases} .$$

The sequence h is commonly called the *impulse response* of the system defined by equation (3.4.1).

In systems theory there are two important ways how the concept of impulse response, or weight function, as it is also called, is used. First, it is closely connected to the so called BIBO stable systems, and second, when functional transform methods, such as the Z – transform, is used, the Z – transform of the impulse response is called the transfer function of the system.

BIBO stable systems can be defined as follows: A system described by equation (3.4.1) is called BIBO stable iff for any input x such that $|x(\alpha)| \leq M$ for all $\alpha \in A$ there is $|y(\gamma)| \leq N$ for all $\gamma \in A + B$, i.e. iff $x \in \ell_\infty \Rightarrow y \in \ell_\infty$. Application of the triangular inequality yields immediately the following result.

Theorem 3.4.7. A linear translation invariant system is BIBO stable iff its impulse response h is absolutely summable, i.e. iff $\sum_{\gamma \in A + B} |h(\gamma)| < \infty$ or $h \in \ell_1$.

Functional transforms are in discrete systems theory introduced in two different ways: either as formal power series or as absolutely convergent series on some subsets of $\mathbb{C}^n$.

First of these approaches is relatively simple, but its use for operations on inputs and outputs of systems is limited – many operations, in particular division – cannot be justified. The other way is more involved, since it demands the use of rather complicated results on convergent power series in n complex variables. Both these approaches have one property in common: Given such transforms P, Q of sequences p, q the transform of the sequence $p * q$ is given by the product PQ. For this reason are the transforms called convolutional. Note that e. g. the one – dimensional Laplace transform or the Fourier transform has under corresponding assumptions the same property.

The straightforward generalization of the one – dimensional systems in the above described form is to admit the countable sets, on which inputs and outputs are defined, to be subsets of $\mathbb{C}^A$, $A \subseteq \mathbb{Z}^n$. Such assumption makes the n – D systems theory very close to partial difference equations, their inputs and outputs to be sequences.

In view of Theorem 3.2.6 the previous concepts and results can be with evident and non – essential changes (among others the absolute value has to be replaced by the mutually corresponding matrix and vector norms) reworded for linear translation invariant systems defined by systems of PDE's with coefficients to be constant square matrices. In the above formulated theorem the δ on the right – hand side of the equation has to be replaced by $\delta \mathbf{I}$, where $\mathbf{I}$ is the unit matrix of corresponding dimension.

Application of these results to various models of linear 2 – D discrete systems which are mostly used is evident mainly because these models are very simple.

The first models of two – dimensional linear discrete systems were presented as systems of PDE's of special form adapted to the introduced notation for these PDE's the set $A = \{(m, n) : m \geq 0, n \geq 0\}$, i. e. here it is the so called first quadrant. Omitting for now explanation of their system – theoretical meaning, the equations have the following form:

1. In the Fornasini – Marchesini model

$$\mathbf{a}(m, n)\mathbf{y}(m + 1, n + 1) + \mathbf{b}(m, n)\mathbf{y}(m + 1, n) +$$
$$+ \mathbf{c}(m, n)\mathbf{y}(m, n + 1) + \mathbf{d}(m, n) = \mathbf{x}(m, n),$$

   where $\mathbf{x}$, $\mathbf{y}$ are d – dimensional vectors, $\mathbf{a}$, $\mathbf{b}$, $\mathbf{c}$, $\mathbf{d}$ are square matrices of dimension d. In most of the applications these matrices are considered to be constant, independent on m and n.

2. The Roesser's model is characterized by the following equations

$$\left[\begin{array}{c} \mathbf{y}^{(h)}(\alpha + (1, 0)) \\ \mathbf{y}^{(v)}(\alpha + (0, 1)) \end{array} \right] = \mathbf{A}(\alpha) \left[\begin{array}{c} \mathbf{y}^{(h)}(\alpha) \\ \mathbf{y}^{(v)}(\alpha) \end{array} \right] + \left[\begin{array}{c} \mathbf{x}^{(h)}(\alpha) \\ \mathbf{x}^{(v)}(\alpha) \end{array} \right].$$

   Here again $\mathbf{A}$ is a square matrix of dimension d, $\mathbf{y}^{(h)}$, $\mathbf{y}^{(v)}$ are vectors of dimension h and v, respectively, with $h + v = d$ and similarly for $\mathbf{x}^{(h)}$, $\mathbf{x}^{(v)}$. Here also the matrix $\mathbf{A}$ is mostly considered to be constant.

3. Attasi suggested a model similar to that of Fornasini – Martesini's with the coefficient $\mathbf{d}(m, n) \equiv 0$.

These models of 2 – D systems are in a certain sense equivalent: any one of them can be matched by any other by rearranging and renaming its coefficients and variables (see [5]).

In all these models application of above presented results is very simple. It is easy to see that the equations have unique and recursively computable solutions with initial values on the set $G = \{(m, n) : m = 0 \ or \ n = 0\}$. Various generalizations of these models with $d > 2$ were also considered with no difficulties. The limitation to A to be $A = \{\alpha = (m, n) : m \geq 0, n \geq 0\}$, and also the assumption $d = 2$ turned out to be too restrictive in applications. With the results described above we may obtain also further results on qualitative properties of solution, which very important in systems theory.

We will consider systems described by a single PDE and consider such equation as an input – output relation, later we will extend these results to systems of linear equations.

One of the basic issues in system theory is the stability of systems [18], [19], [5], [4]. In 2 – D or n – D linear discrete systems mostly the so called BIBO stability is being dealt with. In its common definition some assumptions are made either explicitly or tacitly, which cannot be accepted in our more general setting (see examples below). Moreover, as it is shown in Theorem 3.2.3 and subsequent considerations, the equation (3.2.2) alone (without the specification of the initial set G) does not define a system in a way common in general system theory (see [20], [3], [6]). Equation (3.2.2) together with a specified initial set G defines a linear discrete system and stability of solutions of equation (3.2.2) without such specification cannot be correctly defined. The definition of stability given below is related to the system defined by equation (3.2.2) together with its initial set.

In what follows discrete functions with domain D will belong to the normed space l_p, $p \geq 1$ with the norm defined by $\|y\|_D^p = \sum_{\alpha \in D} |y(\alpha)|^p$. Commonly, the case l_∞ with $\|y\|_D = \sup_{\alpha \in D} |y(\alpha)|$ is also included. Note that for all $p \geq 1$ and for all sequences there is $\sup_{\alpha \in D} |y(\alpha)| \leq \|y\|_D$. Any use of $\|.\|$ assumes that the corresponding domain is given. Relations between norms with different domains of the involved sequences are not excluded.

Definition 3.4.8. The system defined by equation (3.2.2) together with its initial set G will be called *initial state stable* w. r. t. $\|.\|$ iff there exists a positive constant K such that for $x \equiv 0$ and for any bounded initial state $g : G \to \mathbb{C}$ the solution y of (3.2.2) satisfies inequality $\|y\| \leq K\|g\|$.

The system defined by equation (3.2.2) together with its initial set G will be called *input stable* w. r. t. $\|.\|$ iff there exists a positive constant K such that for $g \equiv 0$ and for any bounded input $x : G \to \mathbb{C}$ the solution y of (3.2.2) satisfies inequality $\|y\| \leq K\|x\|$.

It will be convenient to call a system to be initial state or input *stable on the set* $\tilde{A} \subseteq A$ when conditions of the above definition are satisfied on the set $\tilde{A}$. (Precisely speaking, inequalities $\|y\|_{A+B} \leq K\|g\|_G$, $\|y\|_{A+B} \leq K\|x\|_A$ are replaced by $\|y\|_{\tilde{A}+B} \leq$

$K\|g\|_G$, $\|y\|_{\tilde{A}+B} \leq K\|x\|_{\tilde{A}}$.) We shall also use the almost self–evident concept of non–stable systems for those where at least one input or initial state can be found for which no finite constant K ensures the validity of the estimates in the above definition.

Note that in some cases input stable systems are also initial state stable, since the initial values g can be replaced by equivalent inputs. Since in most of the results of this paper the initial state and input stability occur simultaneously, it might be questionable whether two different types of stability must be considered. The following lemma and subsequent comments shows a class of linear difference equations with constant coefficients, which are initial stable but not input stable.

For equation (3.4.1) with constant coefficients the multivariable polynomial $D(z) = \sum_{\beta \in B} \alpha_\beta z^\beta$ with $z \in \mathbb{Z}^n$, $z^\beta = z_1^{\beta_1} z_2^{\beta_2} \ldots z_n^{\beta_n}$ will be called the characteristic polynomial of the system defined by equation (3.4.1).

The concept of the characteristic polynomial D will be used here also as an operator, i. e. any polynomial (possibly in positive and/or negative powers) maps a sequence, e. g. $(D(z)y)(\alpha)$ describes the left–hand side of (3.4.1).

Lemma 3.4.9. Let q denote any solution of the homogeneous difference equation $Qy = 0$, where Q is a given polynomial. Denote $P = z_1 \frac{\partial Q}{\partial z_1}$. Then the sequence $y = \alpha^1 q(\alpha)$ is a solution of the equation $Qy = Pq$.

Proof. The proof follows by substitution. Let $Q(z) = \sum_{\beta \in B} b(\beta)z^\beta$. Then $P(z) = \sum_{\beta \in B} \beta^1 b(\beta)z^\beta$. A substitution of $\alpha^1 q(\alpha)$ for y in Qy yields

$$\sum_{\beta \in B} b(\beta)y(\alpha + \beta) = \sum_{\beta \in B} b(\beta)(\alpha^1 + \beta^1)\, q(\alpha + \beta)$$

$$= \alpha^1 \sum_{\beta \in B} b(\beta)q(\alpha + \beta) + \sum_{\beta \in B} b(\beta)\beta^1\, q(\alpha + \beta).$$

Since q is a solution of the homogeneous equation the first sum equals zero. The second sum equals Pq. $\qquad\square$

The Lemma 3.4.9 is here formulated for the "first" variable α^1 but it evidently holds true for any of the variables and also for their combination e. g. as with $P = \sum_i^n z_i \frac{\partial Q}{\partial z_i}$. Also the operator P can be iterated so as to obtain more complicated patterns of the lemma. This is not the way of reasoning we want to follow here.

Note that the Lemma 3.4.9 does not contain any reference to the domain A of the involved sequences, i. e. it remains true for any $A \subseteq \mathbb{Z}^n$.

The main corollary of this lemma is as follows. When a system defined by a linear equation with constant coefficients and with a given initial set G is initial state stable then its solution y may or may not have to satisfy the requirement $\lim_{\alpha \to \infty} y(\alpha) = 0$. If not, then according to the above lemma, the system cannot be input stable, since there exist unbounded solutions for some special inputs. A similar situation arises e. g. for ordinary difference equations ($n = 1$) when the characteristic polynomial $Q \neq 0$ for all $|z| > 1$ and Q has a zero on the boundary of the unit disk. Here some bounded inputs may have unbounded outputs. Hence the above Lemma 3.4.9 describes an n–D analogy of resonance for $n-D$ systems, which is well known for $1-D$ systems, both continuous and discrete.

Example 3.4.10. Consider the equation $2y(m+1,n+1) = y(m,n+1)+y(m+1,n)$, $m \geq 0, n \geq 0$ with a given bounded function $y : G \to \mathbb{R}$, $G = \{(m,n) : mn = 0\}$. According to Theorem 3.3.1 it has a bounded solution, e. g. for $y = 1$ on the set G we have $y = 1$ for all $m, n \geq 0$. Nevertheless, with $x(m,n) = 1$ for all $m, n \geq 1$ instead of zero at the right-hand side we obtain that $y(m,n) = m$ is an (unbounded) solution. (Note also, that the considered equation satisfies conditions of Theorem 3.3.1, but those of Theorem 3.3.5 are not satisfied.)

For BIBO stability the simplest situation arises when the specific element $\beta_\alpha \in B$, arising in the denominator of the right-hand side of equation (3.2.6) and subsequently called the *leading element* of the recursion, remains unchanged for all $\alpha \in A$. It will be seen that the leading element $\beta_\alpha \in B$ is constant with respect to α (independent of α, so $\beta_\alpha = \beta_0$) e. g. when A is a subset of a (proper) sub-semigroup of $\mathbb{Z}^n$. In this case the following theorem is easy to prove:

Theorem 3.4.11. Let in equation (3.2.2) the initial index set G and the ordering is fixed so that the coefficient $a(\alpha, \beta_0) \neq 0$ for the leading element β_0 and for all $\alpha \in A$. If

$$|a(\alpha, \beta_0)| > \sum_{\beta \in B_0} |a(\alpha, \beta)| \quad \text{for all } \alpha \in A$$

then the system described by (3.2.2) is both input and initial state stable w. r. t. the l_p-norm.

Conditions formulated in this theorem are sufficient but not necessary for systems stability. Most of the efforts so far have been given to equations with constant coefficients, hence they are based on the concept of impulse response, if it is meaningful. Necessary and sufficient conditions for stability have been formulated in terms of the impulse response (3.4.2) and its characteristic function. Results in this direction, originated in the work of Sh. A. Dautov, were summarized in [21]. They are based on rather involved constructions of multivariate analytic functions theory and they hardly can be reproduced here.

It would also be desirable to find conditions of BIBO stability in terms of the coefficients of the equation (3.4.1) or, as it is mostly done, in terms of the characteristic polynomial of this equation. To illustrate this approach we can reproduce here one result on the so called first-quadrant recursive filter stability. Using the simplified notation of this paper, such system is defined by a PDE of the following form

$$\sum_{r=0,s=0}^{r=M,s=N} a_{rs}y(m+r, n+s) = x(m,n), \ m,n \geq 0, \ a_{MN} \neq 0$$

with a slightly modified characteristic polynomial

$$B(z_1, z_2) = z_1^M z_2^N \sum a_{rs} z_1^{-r} z_2^{-s}.$$

Among various results the following one illustrates the situation (see [18]).

Theorem 3.4.12. The system described by the above equation is BIBO stable if and only if

1. $B(z_1, z_2) \neq 0$ for all $|z_1| = |z_2| = 1$,

2. $B(z, z) \neq 0$ for all $|z| \leq 1$.

Since $n-$D discrete systems theory is almost exclusively based on PDE's, we wanted to summarize here some results where these topics overlap. We also wanted to give some impulses for further work in this waste area of investigations.

3.5 Conclusions.

Partial difference equations can be treated using various approaches. In the literature one can find algebraic tools, methods of functional analysis, the use of multivariate analytic function theory, methods of numerical analysis. In this survey we presented a way which can be characterized as a set-theoretical approach, since the basic theorems follow from interconnection of the sets A, B, G, H all belonging to $\mathbb{Z}^n$ as they were denoted in the text. With this approach the existence and unicity theorems for initial value problems and some boundary value problems were formulated for all linear PDE's and a broad class of nonlinear PDE's as well as for systems of such equations. The ordering of the countable sets allowed for some growth estimates and other qualitative properties of solutions. Application of these results to discrete $n-$D systems theory uses also the set-theoretical approach and tries to connect PDE's with general systems theory. Here we deliberately avoided the state space description of systems and also functional transforms often used for description of $n-$D discrete systems and concentrated on input-output relations of these systems in the "time domain".

The most results are presented in this paper without proofs if these proofs are available in the literature. We tried instead to present examples and other supporting material to help the reader to understand the background of results and give some hints for further work. The mathematics of PDE's is far from being complete even in case of linear equations with constant coefficients. We wanted in this paper to contribute to formulation of problems to be solved, since we believe that a proper formulation of a problem is half-way to its solution.

References

[1] R.P. Agarwal. *Difference equations and inequalities. Theory, methods and applications*. Marcel Dekker Inc., 1992.

[2] J. Hekrdla. Index transforms for n–dimensional dft's. *Numerische Mathematik*, 51:469–480, 1987.

[3] U. Oberst. Multidimensional constant linear systems. *Acta Appl Applicandae Mathematicae*, 20:1–175, 1990.

[4] N. K. Bose. *Applied mathematical systems theory*. Van Nostrand, 1982.

[5] T. Kaczorek. *Two-dimensional linear systems*. Springer, Englewood Cliffs, 1985.

[6] P. Rocha. *Structure and representation of 2-D systems*. Univ. Groningen, 1990.

[7] M. Bosk and Gregor J. On generalized difference equations. *Apll. Math. (Prague)*, 32(3):224–239, 1987.

[8] J. Gregor. The cauchy problem for partial difference equations. *Acta Applicandae Mathematicae*, 53:247–263, 1998.

[9] J. Gregor. Singular systems of partial difference equations. *Multidimensional Systems and Signal Processing*, 4:67–82, 1993.

[10] F. A. Gantmacher. *The theory of matrixes*. New York: Chelsea, 1959.

[11] J. Veit. Boundary value problems for partial difference equations. *Multidimensional Systems and Signal Processing*, 7:113–134, 1996.

[12] C. de Boor, Hlling K., and Riemenscheider S. Fundamental solutions of multivariate difference equations. *J. Amer. Math.*, 111:403–415, 1989.

[13] J. Veit. Fundamental solutions of partial difference equations. *Z. Angew. Math. Mech.*, 83(1):51–59, 2003.

[14] F. G. Boese. Assymptotical stability of partial difference equations with variable coefficients. *J. Math. Anal. Appl.*, 276:702–722, 2002.

[15] J. Veit. Sub–exponential solutions of multidimensional difference equations. *Multidimensional Systems and Signal Processing*, 4:369–385, 1997.

[16] J. Gregor. The maximum principle and growth estimates for partial difference equations. *Aequationes Mathematicae*, 71:86–99, 2006.

[17] J. Gregor. Convolutional solutions of partial difference equations. *Mathematics of Control, Signals and systems*, 4:205–216, 1991.

[18] B. T. O'Connor and T. S. Huang. Stability of general two–dimensional recursive filters. *in: Two–Dimensional Digital Signal Processing I, Ed.: T. S. Huang, Springer, Berlin*, 1981.

[19] D. Goodman. Some stability properties of two–dimensional linear shift–invariant digital filters. *IEEE Trans. Circuits and Systems*, 24:201–207, 1977.

[20] M. D. Mesarovic and Y. Takahara. *General systems theory: Mathematical foundations*. Academic Press, New York, 1975.

[21] N. K. Bose. *Multidimensional systems theory*. Reidel, D., 1985.

CHAPTER 4

Numerical schemes and difference equations

Efstratios E. Tzirtzilakis[1] and Nikolaos G. Kafoussias[2]

[1]Department of Mechanical Engineering and Water Resources, Technological Educational Institute of Messolongi, 30200 Messolongi, Greece.

email: tzirtzi@iconography.gr

and

[2]Department of Mathematics, Section of Applied Analysis, University of Patras, 26500 Patras, Greece.

e-mail: nikaf@math.upatras.gr

Abstract: In this study we investigate the connection of difference equations and numerical schemes through the study of a simple partial differential equation (pde). After an indroduction to different numerical schemes, we use some well known finite differences schemes to discretize the simple linear pde $u_t + 2u_x = 0$ with initial condition $u(x,0) = x$. The different discretization schemes lead to different, consistent to the original pde, numerical schemes which constitute corresponding partial difference equations. The solution of the above mentioned pde is attained numerically as well as by analytic solution of the corresponding difference equations. The results show that the solution is always attained by using the analytic solution of difference equations whereas, limitations should be taken into consideration when we try to achieve the solution numerically. These results indicate that the analytic solution of difference equations, resulting from application of numerical schemes, could be of extreme importance for the estimation of the solution of a pde.

Key words and phrases: analytic solution, difference equations, differential equations, basic theory of numerical schemes

4.1   Introduction.

The mathematical formulation of many problems in Mathematical Physics involves rates of change of a quantity, let' s say u, with respect to two or more independent variables. This formulation leads to a partial differential equation (PDE) for the dependent variable u and the conditions that this dependent variable must satisfy round the boundary curve C, of the domain D of its definition, are termed the boundary (and/or initial) conditions.

As special cases of a two dimensional, second-order partial differential equation can be referred:

The Laplace equation $\dfrac{\partial^2 u}{\partial x^2} + \dfrac{\partial^2 u}{\partial y^2} = 0$ (elliptic type),

The Diffusion equation $\dfrac{\partial u}{\partial t} = k\dfrac{\partial^2 u}{\partial x^2}$ (parabolic type) and

The Wave equation $\dfrac{\partial^2 u}{\partial t^2} = c\dfrac{\partial^2 u}{\partial x^2}$ (hyperbolic type)

The above equations become more complicated in three or four dimension problems and to the present, only a limited number of special cases of these equations have been solved analytically. The usefulness of these solutions is further restricted to problems involving shapes for which the boundary conditions can be satisfied. This not eliminates all problems with boundary curves that are undefined in terms of equations, but also many for which the boundary conditions are too difficult to satisfy even though the equations for the boundary curves are known.

On the other hand, many contemporary problems of interest in science and technology are described by a system of non linear and coupled partial differential equations subject to complicated initial and boundary conditions. In such cases approximation methods, whether analytical or numerical in character, are the only means of solution, apart from the use of analogue devices.

Analytical approximation methods often provide extremely useful information concerning the character of the solution for critical values of the dependent variables but tent to be more difficult to apply than the numerical methods. On the contrary, the numerical approximation methods or computational techniques, replace the governing partial differential equation into a **difference equation** or to **a system of algebraic equations**, so that a computer can be used to obtain the solution.

During the last decades the high - speed computing machines has made possible the solution of scientific and engineering problems of great complexity and a lot of numerical methods or computational techniques (**numerical schemes**) have been developed to obtain the solution. As the most well known numerical methods for solving differential equations can be referred the following:

- Finite Element Method (FEM)

- Spectral Methods (SM)

- Finite Volume Method (FVM)

- Finite Difference Method (FDM)

Among the numerical approximation methods, available for solving differential equations, those employing finite-differences are more frequently used and more universally applicable than any other. So, in what follows, we restrict ourselves in presenting the fundamental ideas of finite-difference method, the various numerical schemes that are associated with this method and their connection with differences equation or to a system of algebraic equations. For the three other numerical approximation methods one could say that:

Finite Element Method is frequently used to solve partial differential equations of elliptic type that occur in engineering applications and use piecewise functions (e.g. linear or quadratic), valid on elements of the domain D, to describe the local variations of the unknown function u. The governing equation is precisely satisfied by the exact solution u. If the piecewise approximating functions for u are substituted into the equation it will not hold exactly and a residual is defined to measure the errors. Next the residuals (and hence the errors) are minimized in some sense by multiplying them by a set of weighting functions and integrating. As a result we obtain a set of algebraic equations for the unknown coefficients of the approximating functions. The theory of finite elements has been developed initially for structural stress analysis.

One advantage of the finite–element method over finite–difference methods is the relative ease with which the boundary conditions of the problem are handled. Many physical problems have boundary conditions involving derivatives and, in general, the boundary of the region is irregularly shaped. Boundary conditions of this type are very difficult to handle using finite–difference techniques, since each condition involving a derivative must be approximated by a difference quotient at the grid points, and irregular shaping of the boundary makes placing the grid points difficult. The finite–element method includes the boundary conditions as integrals in a functional that is being minimized, so the basis–construction procedure is independent of the particular boundary conditions of the problem.

Spectral Methods approximate the unknown function by means of truncated Fourier series or series of Chebyshev polynomials. Unlike the finite difference or finite element approach, the approximations are not local but valid throughout the entire computational domain. Again we replace the unknowns in the governing equation by the truncated series. The constraint that leads to the algebraic equations for the coefficients of the Fourier or Chebyshev series is provided by a weighted residuals concept similar to the finite element method or by making the approximate function coincide with the exact solution at a number of grid points. The advantage of this methods is the hight accuracy of the computed solution demanding relatively very short times of computation.

Finite Volume Method was originally developed as a special finite difference formulation and it is also known as **Control Volume Method (CVM)**.

Finite Difference Method is a very general method for solving PDE's. In general, obtaining computational solutions consists of two stages:

(a) The reduction of the partial differential equation to a difference equation or to a system of algebraic equations and

(b) the solution of these equations. The first stage is usually called discretization.

Finite Difference methods are approximate in the sense that derivatives at a point of the domain are approximated by difference quotients over a small interval, i.e., $\frac{\partial u}{\partial x}$ is replaced by $\delta u/\delta x$ where δx is small, but the solutions are not approximate in the sense of being crude estimates. The data of the problems of technology are invariably subject to errors of measurement, besides which, all arithmetical work is limited to a finite number of significant figures and contains rounding errors, so even analytical solutions provide only approximate numerical answers. Finite Difference methods generally give solutions that are either as accurate as the data warrant or as accurate as is necessary for the technical purposes for which the solutions are required. In both cases a finite difference solution is as satisfactory as one calculated from an analytical formula.

There is no difficulty in formally applying finite difference methods to non–linear differential equations. The difficulties are associated with the **difference equations** or with **the system of algebraic equations** themselves. If they are linear they can usually be solved quite easily, although we still have the problem of determining the conditions that must be satisfied for stability and convergence because the coefficients of the unknowns will be functions of the solution at earlier time–levels. If they are non–linear we have also the problem of their solution. **Direct methods**, in general, are difficult, so they are usually solved **iteratively** after being linearized in some way.

It must be also emphasized that a good understanding of the numerical solution method is very crucial. Three mathematical concepts are implicated in determining the success of such a method or technique: consistency, stability and convergence.

Consistent numerical schemes give rise to systems of algebraic equations, which can be demonstrated to be equivalent to the original governing equation as the grid spacing tends to zero.

Stability is associated with damping of errors as the numerical method proceeds. If a technique is not stable even round off errors in the initial data can cause wild oscillations or divergence.

Convergence is the property of a numerical method to produce a solution which approaches the exact solution as the grid spacing, control volume size or element size is reduced to zero.

Convergence is usually very difficult to establish theoretically and in practice the **Lax's equivalence theorem** is used which states that for linear problems a necessary and sufficient condition for convergence is that the method must be both consistent and stable. However, in many cases this theorem is of limited use since the governing equations are non–linear. In such problems consistency and stability are necessary conditions for convergence, but not sufficient.

Descriptive or Analytical Treatment of the Consistency, Stability and Convergence of Finite Difference Numerical Schemes

The solution of the finite difference equation, used to approximate the partial differential equation, is a reasonably accurate approximation to the solution of the corresponding partial differential equation, when some conditions are satisfied. These conditions are associated with **two different** but interrelated problems. The first concerns the convergence of the exact solution of the approximating difference equations to the solution of the partial differential equation. The **second** one, concerns the unbounded growth or controlled decay of any kind of errors associated with the solution of the finite difference equations.

Descriptive Treatment of Convergence

Let $\bar{u}$ represent the exact solution of a partial differential equation with independent variables x and t, and u the exact solution of the difference equations used to approximate the partial differential equation. Then the finite-difference equation is said to be convergent when u tends to $\bar{u}$ at a fixed point or along a fixed t-level as δx and δt both tend to zero.

Although the conditions under which u converges to $\bar{u}$ have been established for linear elliptic, parabolic and hyperbolic second–order partial differential equations, with solutions satisfying fairly general boundary and initial conditions, they are not yet known for non–linear equations except in a few particular cases. In general the problem of convergence is a difficult one to investigate usefully, because the final expression for the discretization error (or truncation error $\bar{u}(x,t) - u(x,t)$) is usually in terms of unknown derivatives for which no bounds can be estimated. Fortunately however, the convergence of difference equations approximating linear parabolic and hyperbolic differential equations can be investigated in terms of stability and consistency which are much easier to deal with.

Analytical Treatment of Stability

There are two standard ways of investigating the boundedness of the solution of the finite–difference equations. In one the equations are expressed in matrix form and the eigenvalues are examined of an associated matrix. In the other, a finite Fourier series is used. The Fourier method is the easier of the two in that it requires no knowledge of matrix algebra but is the less rigorous because it neglects the boundary conditions.

Stability by the Fourier Series Method (von Neumann's method)

This method, developed by von Neumann during World War II, was first discussed in detail by O'Brien, Hyman and Kaplan in a paper published in 1951. It expresses an initial line of errors in terms of a finite Fourier series, and considers the growth of a function that reduces to this series for $t = 0$ by a "variables separable" method identical with that commonly used for deriving analytical solutions of partial differential equations. The Fourier series can be formulated in terms of sines or cosines but the algebra is easier if the complex exponential form is used i.e., with $\Sigma a_n cos(\eta\pi x/l))$ or $\Sigma b_n sin(\eta\pi x/l)$ replaced by the equivalent $\Sigma A_n e^{in\pi x/l}$ where $i = \sqrt{-1}$ and l is the interval throughout which the function is defined. Clearly, we need to change the usual notation $u_{i,j}$ to $u(ph, qk) = u_{p,q}$. In terms of this notation $A_n e^{in\pi x/l} = A_n e^{in\pi qh/Nh} = A_n e^{i\beta_n ph}$, where $\beta_n = n\pi/Nh$ and $Nh = l$.

In the next paragraphs, examples are given illustrating the above presented ideas regarding the finite-difference method, employing some representative numerical schemes and the corresponding difference equations or the system of algebraic equations.

4.2   Finite differences

As we mentioned in the Introduction section, the basic idea of the Finite Differences is the transformation of a continuum problem (governing equations along with boundary conditions) to a discrete system which can be solved by using computers. In this paragraph we will briefly demonstrate the finite difference method through a specific example. Let us consider a function of two variables, space and time, $u(x,t)$ defined in a domain $(x,t) : [0,1] \times [0, t_\infty]$. Where t_∞ means that we can choose each time a specific value of t which can be as big as we wish, but in any case finite.

Let's consider now the following simple problem

$$u_t + 2u_x = 0 \text{ with initial condition } u(x,0) = x \qquad (4.2.1)$$

the solution of which is

$$u(x,t) = x - 2t \qquad (4.2.2)$$

where, u_s means partial differentiation with respect to s. i.e. $u_x = \dfrac{\partial u}{\partial x}$, $u_t = \dfrac{\partial u}{\partial t}$, $u_{xx} = \dfrac{\partial^2 u}{\partial x^2}$ and so on.

In order to apply the finite differences method for the solution of this problem we discretize the domain $[0,1] \times [0, t_\infty]$ using a grid towards the x and t direction, as shown in Figure 4.1. On the x direction we use M grid points which define on the x axis the points x_1, x_2, $\cdots, x_M$, whereas, on the t direction we use N grid points which define on the t axis the points t_1, t_2, $\cdots, t_N$. The grid spacing on the x direction is Δx whereas on the t direction is Δt.

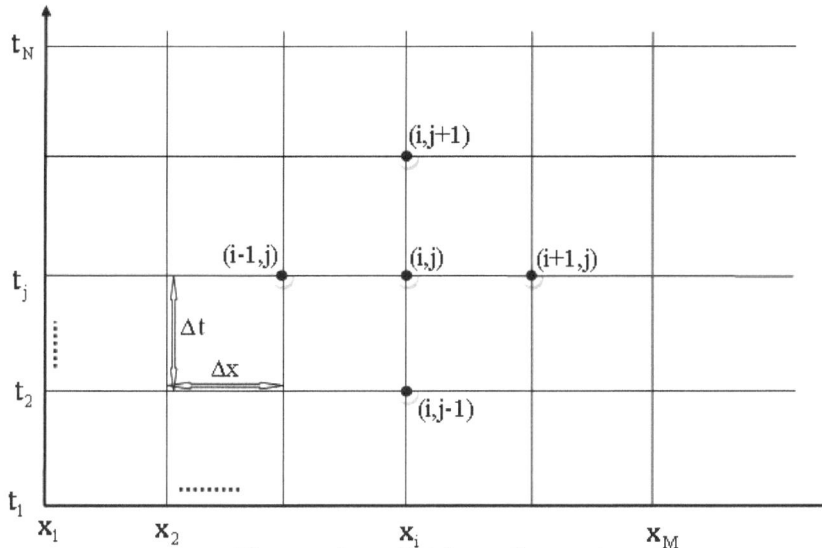

Figure 4.1: Grid configuration

The function $u(x,t)$ in a point (x_0, t_0) of the continuum domain is defined as $u(x_0, t_0)$. The corresponding definition on the discretized grid is $u(i,j)$ with $1 < i < M$ and $1 < j < N$. For simplicity we write

$$u(x_0, t_0) = u(i,j) = u_{i,j} = u_i^j \qquad (4.2.3)$$

We can also define the neighbouring points as follows

$$u(x_0 + \Delta x, t_0) = u_{i+1}^j, \quad u(x_0 - \Delta x, t_0) = u_{i-1}^j$$

$$u(x_0, t_0 + \Delta t) = u_i^{j+1}, \quad u(x_0, t_0 - \Delta t) = u_i^{j-1}$$

The approximation and the solution of a problem using Finite Differences can be demonstrated through expansions of functions with Taylor series. For the point $u(x_0 + \Delta x, t_0)$ we have

$$u\left(x_0 + \Delta x, t_0\right) = u\left(x_0, t_0\right) + \Delta x \left.\frac{\partial u}{\partial x}\right|_{(x_0, t_0)} + \frac{(\Delta x)^2}{2!} \left.\frac{\partial^2 u}{\partial x^2}\right|_{(x_0, t_0)} + \cdots +$$
$$+ \frac{(\Delta x)^{n-1}}{(n-1)!} \left.\frac{\partial^{n-1} u}{\partial x^{n-1}}\right|_{(x_0, t_0)} + \frac{(\Delta x)^n}{n!} \left.\frac{\partial^n u}{\partial x^n}\right|_{(\xi, t_0)} \qquad (4.2.4)$$

where $x_0 \leq \xi \leq x_0 + \Delta x$.

From the above relation we have

$$\left.\frac{\partial u}{\partial x}\right|_{(x_0, t_0)} = \frac{u\left(x_0 + \Delta x, t_0\right) - u\left(x_0, t_0\right)}{\Delta x} - \frac{(\Delta x)}{2!} \left.\frac{\partial^2 u}{\partial x^2}\right|_{(x_0, t_0)} - \cdots \qquad (4.2.5)$$

Using the discretization described above, we can also write that with respect to the (i, j) point. So, we have

$$\frac{\partial u}{\partial x}\bigg|_{(i,j)} = \frac{u(i+1, j) - u(i, j)}{\Delta x} + Truncation\ Error$$

As Truncation Error we consider all the terms neglected from equation (4.2.5). The truncation error is actually the difference of the value of u_x at the point (x_0, y_0) and it's corresponding approximation by the evaluation of u_x using finite differences at the point (i, j). The limiting behaviour of the truncation error is expressed by the symbol "O" and for this specific case we write

$$\frac{\partial u}{\partial x}\bigg|_{(i,j)} = \frac{u(i+1, j) - u(i, j)}{\Delta x} + O(\Delta x) \approx \frac{u(i+1, j) - u(i, j)}{\Delta x} \qquad (4.2.6)$$

where the truncation error is $O(\Delta x)$ which means that

$$|Truncation\ error| \leqslant k\,|\Delta x| \ \ \text{for} \ \Delta x \to 0 \ \text{and} \ k > 0 \ \text{constant}$$

We also say that the approximation of the derivative in equation (4.2.6) is made using "Forward Differences" due to the appearing difference $u(i+1, j) - u(i, j)$. The symbol $\approx$ means approximation.

The above approximation of u_x is not unique . We can consider, for example, the Taylor expansion at the point $u(x_0 - \Delta x, y_0)$ which is

$$u(x_0 - \Delta x, t_0) = u(x_0, t_0) - \Delta x \frac{\partial u}{\partial x}\bigg|_{(x_0,t_0)} + \frac{(\Delta x)^2}{2!} \frac{\partial^2 u}{\partial x^2}\bigg|_{(x_0,t_0)}$$
$$- \frac{(\Delta x)^3}{3!} \frac{\partial^3 u}{\partial x^3}\bigg|_{(x_0,t_0)} + \cdots \qquad (4.2.7)$$

Following the same considerations as above, we produce the approximation of u_x with "Backward Differences" as follows

$$\frac{\partial u}{\partial x}\bigg|_{(i,j)} = \frac{u(i, j) - u(i-1, j)}{\Delta x} + O(\Delta x) \approx \frac{u(i, j) - u(i-1, j)}{\Delta x} \qquad (4.2.8)$$

Moreover, by subtracting (4.2.7) from (4.2.4) we derive the central finite differences approximation for u_x, which is

$$\frac{\partial u}{\partial x}\bigg|_{(i,j)} = \frac{u(i+1, j) - u(i-1, j)}{2\Delta x} + O(\Delta x)^2 \approx \frac{u(i+1, j) - u(i-1, j)}{2\Delta x} \qquad (4.2.9)$$

This latter scheme has truncation error of second order which means that it is more accurate than the other two. This may lead to the false conclusion that by using this scheme we can always produce more accurate and better numerical techniques. On the contrary, we will see that the use of this scheme may lead to an unstable numerical scheme.

Finally, there is a problem with the calculation of the derivative at the boundary points x_1 and x_M as it can also be observed from figure 4.1. For these points and following the above philosophy we can derive the following one-sided differences

$$\frac{\partial u}{\partial x}\bigg|_{(1,j)} = \frac{1}{2\Delta x} \left(-3u(1,j) + 4u(2,j) - u(3,j)\right) + O\left(\Delta x\right)^2$$

$$\frac{\partial u}{\partial x}\bigg|_{(M,j)} = \frac{1}{2\Delta x} \left(3u(M,j) - 4u(M-1,j) + u(M-2,j)\right) + O\left(\Delta x\right)^2$$

Alternatively, for the calculation of the derivatives at the boundary points, we can use forward differences for the point x_1 and backward differences for the point x_M.

Consistency and Stability

Consider the partial differential equation (4.2.1) in which we replace it's derivatives by a finite differences scheme. First we choose to "replace" u_t with forward differences and u_x with backward differences. From the corresponding Taylor expansions we have

$$\frac{\partial u}{\partial t}\bigg|_{(x_0,t_0)} = \frac{u\left(x_0, t_0 + \Delta t\right) - u\left(x_0, t_0\right)}{\Delta t} - \frac{(\Delta t)}{2!}\frac{\partial^2 u}{\partial t^2}\bigg|_{(x_0,t_0)} - \cdots$$

$$\frac{\partial u}{\partial x}\bigg|_{(x_0,t_0)} = \frac{u\left(x_0, t_0\right) - u\left(x_0 - \Delta x, t_0\right)}{\Delta x} + \frac{(\Delta x)}{2!}\frac{\partial^2 u}{\partial x^2}\bigg|_{(x_0,t_0)} - \cdots$$

Consequently (4.2.1) gives,

$$\frac{u\left(x_0, t_0 + \Delta t\right) - u\left(x_0, t_0\right)}{\Delta t} + 2\frac{u\left(x_0, t_0\right) - u\left(x_0 - \Delta x, t_0\right)}{\Delta x} =$$
$$= \frac{(\Delta t)}{2!}\frac{\partial^2 u}{\partial t^2}\bigg|_{(x_0,t_0)} - \Delta x\frac{\partial^2 u}{\partial x^2}\bigg|_{(x_0,t_0)} + \ldots$$

or

$$\frac{u\left(i, j+1\right) - u\left(i, j\right)}{\Delta t} + 2\frac{u\left(i, j\right) - u\left(i-1, j\right)}{\Delta x} -$$
$$- \frac{(\Delta t)}{2!}\frac{\partial^2 u}{\partial t^2}\bigg|_{(x_0,t_0)} + \Delta x\frac{\partial^2 u}{\partial x^2}\bigg|_{(x_0,t_0)} + \ldots = 0$$

or

$$\frac{u\left(i, j+1\right) - u\left(i, j\right)}{\Delta t} + 2\frac{u\left(i, j\right) - u\left(i-1, j\right)}{\Delta x} + O\left(\Delta t\right) + O\left(\Delta x\right) = 0 \qquad (4.2.10)$$

The truncation error for this scheme is

$$Tr.Err. = -\frac{(\Delta t)}{2!}\frac{\partial^2 u}{\partial t^2}\bigg|_{(x_0,t_0)} + \Delta x\frac{\partial^2 u}{\partial x^2}\bigg|_{(x_0,t_0)}$$

and the accuracy is $O\left(\Delta t\right) + O\left(\Delta x\right)$.

Consistent is a numerical scheme of finite differences if

$$\lim_{\Delta x,\, \Delta y \to 0}(Tr.Err.) = 0$$

Clearly, if a numerical scheme is consistent, the computed solution using the numerical scheme, tends to the true (analytic) solution of the original partial differential equation as the grid is constantly refined, i.e. when $\Delta x \to 0$, $\Delta t \to 0$ simultaneously.

From equation (4.2.10), by omitting the terms of accuracy $(O(\Delta x), O(\Delta t))$, it is also obtained that

$$u\left(i, j+1\right) = u\left(i, j\right) - 2\frac{\Delta t}{\Delta x}\left[u\left(i, j\right) - u\left(i-1, j\right)\right]$$

or

$$u_i^{j+1} = u_i^j - 2\frac{\Delta t}{\Delta x}\left(u_i^j - u_{i-1}^j\right) \tag{4.2.11}$$

since the time counter j is used to be written as superscript in the numerical schemes, i.e. $u(i,j) \equiv u_i^j$.

This scheme is called Forward Time Backward Space (FTBS) scheme due to the differences we used for the discretization.

We observe that the above scheme is an explicit "marching problem" since we can calculate the unknown function u at a point x_i at the time step $j+1$ considering known the function u at the previous time step j.

Apart from the approximation error due to the discretization used, there are also other type of errors arising when we compute the solution with a marching scheme. These errors are the induced round–off errors which rise due to the limited ability of a PC to store exactly (or by infinite decimal places) a calculated number. This round–off errors are generally very small. However, if the solution is calculated by an iterative technique or by a marching scheme it is possible these errors to increase additively and create errors known as "accumulation errors" which can lead to "induced instability".

For the problems where we compute the solution through an iterative procedure or in successive time steps (marching problem), we call the numerical scheme **stable** if the arising errors do not increase between two consecutive steps of calculation.

The stability and consistency of a numerical scheme are of vital importance for the convergence of a numerically computed solution. This importance is expressed by the **Lax equivalence theorem**. According to this theorem, for a well posed initial value problem, described by a linear partial differential equation, **consistency and stability is a necessary and sufficient condition for the convergence** of the computed solution to the true solution of the partial differential equation.

4.3   Stability analysis

"Stability" is a common requirement for all the numerical schemes we develop. We can call a numerical scheme "stable", if a creation of an "error" or perturbation, generated at any (random) time of the computational procedure, does not have as a result it's propagation and enlargement at the forthcoming computations of the numerical procedure. The estimation of the accumulation errors, created by the round off errors, is one of the most important problems to deal with. For it's estimation, several techniques have been developed. For a satisfactory large number of partial differential equations the study of stability is attained by the use of Fourier analysis introduced by Von Neumann as already mentioned in the Introduction.

Let $\bar{u}_i^j$ be the computed solution estimated by use of a numerical technique at the point i at the j–th time point. Let also u_i^j the real discrete solution, i.e. the solution we could estimate if the computer had the ability of handling infinite decimal places. If we consider as ε_i^j the error at the same point (i, j) then we can write for every node

$$u_i^j = \bar{u}_i^j + \varepsilon_i^j \tag{4.3.1}$$

As stable, we define a numerical scheme such that

$$\left| \frac{\varepsilon_i^{j+1}}{\varepsilon_i^j} \right| = |G| \leq 1 \quad \text{as} \quad t \to \infty \tag{4.3.2}$$

The ratio G is called amplification factor and as we can easily observe, the above relation means that for a stable numerical scheme the arising errors either remain of the same amplitude ($= 1$) or reduce (< 1) from one iteration (j) to the next ($j + 1$).

We will demonstrate the first steps of the method in the following diffusion equation

$$\frac{\partial u}{\partial t} = a \frac{\partial^2 u}{\partial x^2} \tag{4.3.3}$$

Using forward differences for time and central differences for space, as we described above, we have the following discretization scheme

$$\frac{1}{(\Delta t)} \left(\bar{u}_i^{j+1} - \bar{u}_i^j \right) = \frac{a}{(\Delta x)^2} \left(\bar{u}_{i+1}^j - 2\bar{u}_i^j + \bar{u}_{i-1}^j \right) \tag{4.3.4}$$

We assume that the above equation is satisfied for the computed solution $\bar{u}_n^n$ and for the corresponding "real" solution u_i^j. By using equation (4.3.1) we have

$$\frac{1}{(\Delta t)} \left(u_i^{j+1} - \varepsilon_i^{j+1} - u_i^j + \varepsilon_i^j \right) = \frac{a}{(\Delta x)^2} \left(u_{i+1}^j - \varepsilon_{i+1}^j - 2u_i^j + 2\varepsilon_i^j + u_{i-1}^j - \varepsilon_{i-1}^j \right) \Leftrightarrow$$

$$\frac{1}{(\Delta t)} \left(u_i^{j+1} - u_i^j \right) + \frac{1}{(\Delta t)} \left(-\varepsilon_i^{j+1} + \varepsilon_i^j \right) = \frac{a}{(\Delta x)^2} \left(u_{i+1}^j - 2u_i^j + u_{i-1}^j \right)$$

$$+ \frac{a}{(\Delta x)^2} \left(-\varepsilon_{i+1}^j + 2\varepsilon_i^j - \varepsilon_{i-1}^j \right)$$

and since equation (4.3.4) is also satisfied for u_i^j we conclude to the following equation for the errors ε_i^j:

$$\frac{1}{(\Delta t)}\left(\varepsilon_i^{j+1} - \varepsilon_i^j\right) = \frac{a}{(\Delta x)^2}\left(\varepsilon_{i+1}^j - 2\varepsilon_i^j + \varepsilon_{i-1}^j\right) \tag{4.3.5}$$

We observe that the equation of errors is of the same form with the original discrete equation (4.3.4) and this is generally true for linear equations. According to Von Neumann analysis, errors are introduced in each computational step and propagate in time (as j increases) according to the above equation (4.3.5). Without loss of generality we can assume that the errors are introduced at $t = 0$. Let $E(x_i)$ be the errors introduced at a random position x_i. A harmonic analysis of the errors, by use of a Fourier series, gives

$$E(x_i) = \sum_{k=1}^{M} A_k e^{\mathbb{I}\beta_k x_i} \quad i = 1, 2, \cdots, M \tag{4.3.6}$$

where $\mathbb{I}$ is the complex unit number and $\beta_k = \frac{\Delta x(k-1)\pi}{x_M - x_1}$.

Due to the linearity of (4.3.3), we can use the principle of superposition of solutions. Consequently, from the series (4.3.6) we can consider only one term, $e^{\mathbb{I}\beta x}$, with $x = x_i$, $i = 2, 3, \cdots, M-1$ and β one of the numbers β_k. We seek for a solution for (4.3.5) which reduces to $e^{\mathbb{I}\beta x}$ for $t = 0$. Due to the ability to apply separation of variables method to equation (4.3.3), a unique solution can be found to be of the form

$$\varepsilon_i^j = \varepsilon(x_i, t_j) = \varepsilon(x, t) = e^{\gamma t}e^{\mathbb{I}\beta x} \tag{4.3.7}$$

where γ can be real or complex number.

The form (4.3.7) of the error solution constitutes a general form which can be used for the stability analysis of any numerical scheme of finite differences for one dimensional space problems propagating in time (marching). It is emphasised, that the Von Neumann stability analysis is strictly applied to linear partial differential equations and only for initial value problems with periodic initial conditions. However, this method is very famous and it is excessively applied to non-linear partial differential equations due to luck of corresponding general theorems for these equations. We will demonstrate the application of the Von Neumann stability analysis for equation (4.2.1), for different discretization schemes, in the following subsections.

FTBS

As we already described in the previous session, if we discretize (4.2.1) using Forward differences for Time and Backward for Space, we conclude to the following scheme (FTBS) of order $O(\Delta t) + O(\Delta x)$

$$u_i^{j+1} = u_i^j - 2\frac{\Delta t}{\Delta x}\left(u_i^j - u_{i-1}^j\right) \tag{4.3.8}$$

The corresponding equation of errors, will follow the above equation as we demonstrated in the previous session, and it is

$$\varepsilon_i^{j+1} = \varepsilon_i^j - 2\frac{\Delta t}{\Delta x}\left(\varepsilon_i^j - \varepsilon_{i-1}^j\right) \tag{4.3.9}$$

Keeping in mind that

$$\varepsilon_i^j = e^{\gamma t}e^{\mathbb{I}\beta x} \qquad \varepsilon_i^{j\pm 1} = e^{\gamma(t\pm\Delta t)}e^{\mathbb{I}\beta x} \qquad \varepsilon_{i\pm 1}^j = e^{\gamma t}e^{\mathbb{I}\beta(x\pm\Delta x)}$$

we have

$$e^{\gamma(t+\Delta t)}e^{\mathbb{I}\beta x} = e^{\gamma t}e^{\mathbb{I}\beta x} - 2\frac{\Delta t}{\Delta x}\left(e^{\gamma t}e^{\mathbb{I}\beta x} - e^{\gamma t}e^{\mathbb{I}\beta x}e^{-\mathbb{I}\beta\Delta x}\right) \Longleftrightarrow$$

$$\frac{e^{\gamma(t+\Delta t)}e^{\mathbb{I}\beta x}}{e^{\gamma t}e^{\mathbb{I}\beta x}} = 1 - 2\frac{\Delta t}{\Delta x}\left(\frac{e^{\gamma t}e^{\mathbb{I}\beta x} - e^{\gamma t}e^{\mathbb{I}\beta x}e^{-\mathbb{I}\beta\Delta x}}{e^{\gamma t}e^{\mathbb{I}\beta x}}\right) \Longleftrightarrow$$

$$\frac{\varepsilon_i^{j+1}}{\varepsilon_i^j} = 1 - 2\frac{\Delta t}{\Delta x}\left(1 - e^{-\mathbb{I}\beta\Delta x}\right) \Leftrightarrow G = 1 - 2\lambda\left(1 - e^{-\mathbb{I}\beta\Delta x}\right) \Longleftrightarrow$$

$$G = 1 - 2\lambda\left(1 - \cos\left(\beta\Delta x\right) + \mathbb{I}\sin\left(\beta\Delta x\right)\right) = 1 - 2\lambda + 2\lambda\cos\left(\beta\Delta x\right) - 2\lambda\mathbb{I}\sin\left(\beta\Delta x\right)$$

where $\lambda = \Delta t/\Delta x$ and in this case, the amplification factor G is turned out to be a complex number and we require it's magnitude to be less or equal to 1:

$$|G|^2 \le 1 \quad \Leftrightarrow \quad |G|^2 = \left[1 - 2\lambda + 2\lambda\cos\left(\beta\Delta x\right)\right]^2 + 4\lambda^2\sin^2\left(\beta\Delta x\right) \le 1 \Longleftrightarrow$$

$$1 - 4\lambda + 8\lambda^2 + \left(4\lambda - 8\lambda^2\right)\cos\left(\beta\Delta x\right) \le 1$$

By using the known trigonometric identity

$$\cos\left(\beta\Delta x\right) = 1 - 2\sin^2\left(\frac{\beta\Delta x}{2}\right) \tag{4.3.10}$$

we derive that

$$1 - 8\lambda\left(1 - 2\lambda\right)\sin^2\left(\frac{\beta\Delta x}{2}\right) \le 1$$

Thus

$$|G|^2 \le 1 \quad \Leftrightarrow \quad 1 - 8\lambda + 16\lambda^2 \le 1 \quad \Leftrightarrow \quad 0 \le \lambda \le \frac{1}{2}$$

Consequently, this scheme is what we call "conditionally stable" and the FTBS is stable if $0 < \lambda \le \frac{1}{2}$ the FTBS is stable.

FTCS

By discretization of (4.2.1) using the forward differences for the time derivative and central differences for the space derivative (Forward Time Central Space (FTCS) scheme) we have the following scheme of order $O(\Delta t) + O((\Delta x)^2)$

$$u_i^{j+1} = u_i^j - \frac{\Delta t}{\Delta x}\left(u_{i+1}^j - u_{i-1}^j\right) \tag{4.3.11}$$

The corresponding equation of the errors are

$$\varepsilon_i^{j+1} = \varepsilon_i^j - \frac{\Delta t}{\Delta x}\left(\varepsilon_{i+1}^j - \varepsilon_{i-1}^j\right) \tag{4.3.12}$$

The stability analysis in this case gives

$$e^{\gamma(t+\Delta t)}e^{\mathbb{I}\beta x} = e^{\gamma t}e^{\mathbb{I}\beta x} - \frac{\Delta t}{\Delta x}\left(e^{\gamma t}e^{\mathbb{I}\beta(x+\Delta x)} - e^{\gamma t}e^{\mathbb{I}\beta(x-\Delta x)}\right) \Longleftrightarrow$$

$$\frac{e^{\gamma(t+\Delta t)}e^{\mathbb{I}\beta x}}{e^{\gamma t}e^{\mathbb{I}\beta x}} = 1 - \frac{\Delta t}{\Delta x}\left(\frac{e^{\gamma t}e^{\mathbb{I}\beta x}e^{\mathbb{I}\beta\Delta x} - e^{\gamma t}e^{\mathbb{I}\beta x}e^{-\mathbb{I}\beta\Delta x}}{e^{\gamma t}e^{\mathbb{I}\beta x}}\right) \Longleftrightarrow$$

$$\frac{\varepsilon_i^{j+1}}{\varepsilon_i^j} = 1 - \frac{\Delta t}{\Delta x}\left(e^{\mathbb{I}\beta\Delta x} - e^{-\mathbb{I}\beta\Delta x}\right) \Rightarrow G = 1 - \lambda\left(e^{\mathbb{I}\beta\Delta x} - e^{-\mathbb{I}\beta\Delta x}\right)$$

Thus, the amplification factor takes finally the form

$$G = 1 - \lambda\left(cos\left(\beta\Delta x\right) + \mathbb{I}\sin\left(\beta\Delta x\right) - cos\left(\beta\Delta x\right) + \mathbb{I}\sin\left(\beta\Delta x\right)\right) =$$

$$= 1 - 2\lambda\mathbb{I}\sin\left(\beta\Delta x\right)$$

Since G is a complex number we require it's magnitude to be less or equal to 1. However, it is

$$|G|^2 = 1 + 4\lambda^2\sin^2\left(\beta\Delta x\right) \geq 1$$

and thus the FTCS scheme is unconditionally unstable. At this point we note that the use of this more accurate difference scheme, than the FTBS which is $O(\Delta x)$ instead of $O((\Delta x)^2)$ of the present one, finally results ta an unstable scheme.

FTFS

By discretization of (4.2.1) using the Forward Time Forward Space scheme we arrive at the following (FTFS) numerical scheme of order $O(\Delta t) + O(\Delta x)$

$$u_i^{j+1} = u_i^j - 2\frac{\Delta t}{\Delta x}\left(u_{i+1}^j - u_i^j\right) \tag{4.3.13}$$

The corresponding equation of errors is

$$\varepsilon_i^{j+1} = \varepsilon_i^j - 2\frac{\Delta t}{\Delta x}\left(\varepsilon_{i+1}^j - \varepsilon_i^j\right) \Longleftrightarrow \tag{4.3.14}$$

The stability analysis in this case gives

$$e^{\gamma(t+\Delta t)}e^{\mathbb{I}\beta x} = e^{\gamma t}e^{\mathbb{I}\beta x} - 2\frac{\Delta t}{\Delta x}\left(e^{\gamma t}e^{\mathbb{I}\beta(x+\Delta x)} - e^{\gamma t}e^{\mathbb{I}\beta x}\right) \Longleftrightarrow$$

$$\frac{e^{\gamma(t+\Delta t)}e^{\mathbb{I}\beta x}}{e^{\gamma t}e^{\mathbb{I}\beta x}} = 1 - 2\frac{\Delta t}{\Delta x}\left(\frac{e^{\gamma t}e^{\mathbb{I}\beta x}e^{\mathbb{I}\beta \Delta x} - e^{\gamma t}e^{\mathbb{I}\beta x}}{e^{\gamma t}e^{\mathbb{I}\beta x}}\right) \Longleftrightarrow$$

$$\frac{\varepsilon_i^{j+1}}{\varepsilon_i^j} = 1 - 2\frac{\Delta t}{\Delta x}\left(e^{\mathbb{I}\beta \Delta x} - 1\right) \Rightarrow G = 1 - 2\lambda\left(e^{\mathbb{I}\beta \Delta x} - 1\right) \Longleftrightarrow$$

$$G = 1 - 2\lambda\left(\cos\left(\beta\Delta x\right) + \mathbb{I}\sin\left(\beta\Delta x\right) - 1\right) = 1 + 2\lambda - 2\lambda\cos\left(\beta\Delta x\right) - 2\lambda\mathbb{I}\sin\left(\beta\Delta x\right)$$

The magnitude of the amplification factor is

$$|G|^2 = \left[1 + 2\lambda - 2\lambda\cos\left(\beta\Delta x\right)\right]^2 + 4\lambda^2\sin^2\left(\beta\Delta x\right) =$$

$$1 + 4\lambda + 8\lambda^2 - 4\lambda\left(1 + 2\lambda\right)\cos\left(\beta\Delta x\right) \stackrel{(4.3.10)}{=} 1 + 4\lambda\left(1 + 2\lambda\right)\sin^2\left(\frac{\beta\Delta x}{2}\right)$$

Consequently,

$$|G|^2 \leq 1 \Leftrightarrow 1 + 4\lambda\left(1 + 2\lambda\right) \leq 1$$

which is impossible and thus the FTFS scheme is unconditionally unstable.

Leap–Frog

Finally, we use central differences for the time derivative and forward differences for the spatial discretization. The schemes using central differences for time are known as Leap–Frog schemes because the u_i^{j+1} is calculated with a "leap" using also u_i^{j-1}. This scheme mentioned from know on as Leap Frog with forward space (LFFS) differences, is of order $O((\Delta t)^2) + O(\Delta x)$ and gives the following equation

$$u_i^{j+1} = u_i^{j-1} - 4\frac{\Delta t}{\Delta x}\left(u_{i+1}^j - u_i^j\right) \tag{4.3.15}$$

The corresponding equation of errors is

$$\varepsilon_i^{j+1} = \varepsilon_i^{j-1} - 4\frac{\Delta t}{\Delta x}\left(\varepsilon_{i+1}^j - \varepsilon_i^j\right) \tag{4.3.16}$$

The stability analysis in this case gives

$$e^{\gamma(t+\Delta t)}e^{\mathbb{I}\beta x} = e^{\gamma(t-\Delta t)}e^{\mathbb{I}\beta x} - 4\frac{\Delta t}{\Delta x}\left(e^{\gamma t}e^{\mathbb{I}\beta(x+\Delta x)} - e^{\gamma t}e^{\mathbb{I}\beta x}\right) \Longleftrightarrow$$

$$\frac{e^{\gamma(t+\Delta t)}e^{\mathbb{I}\beta x}}{e^{\gamma t}e^{\mathbb{I}\beta x}} = \frac{e^{\gamma t}e^{-\gamma\Delta t}e^{\mathbb{I}\beta x}}{e^{\gamma t}e^{\mathbb{I}\beta x}} - 4\frac{\Delta t}{\Delta x}\left(\frac{e^{\gamma t}e^{\mathbb{I}\beta x}e^{\mathbb{I}\beta\Delta x} - e^{\gamma t}e^{\mathbb{I}\beta x}}{e^{\gamma t}e^{\mathbb{I}\beta x}}\right) \Longleftrightarrow$$

$$\frac{\varepsilon_i^{j+1}}{\varepsilon_i^j} = e^{-\gamma\Delta t} - 4\frac{\Delta t}{\Delta x}\left(e^{\mathbb{I}\beta\Delta x} - 1\right)$$

Also

$$e^{-\gamma\Delta t} = e^{\gamma t - \gamma t - \gamma\Delta t} = \frac{e^{\gamma t}}{e^{\gamma t}e^{\gamma\Delta t}} = \frac{e^{\gamma t}e^{\mathbb{I}\beta x}}{e^{\gamma(t+\Delta t)}e^{\mathbb{I}\beta x}} = \frac{\varepsilon_i^j}{\varepsilon_i^{j+1}} = \frac{1}{G}$$

and thus the amplification factor is turned out to be

$$G = \frac{1}{G} - 4\lambda\left(e^{\mathbb{I}\beta\Delta x} - 1\right) \Rightarrow G^2 + 4\lambda\left(e^{\mathbb{I}\beta\Delta x} - 1\right)G - 1 = 0$$

After some manipulation we calculate the roots of the above equation and thereafter, the corresponding magnitudes. It is turned out that $|G| > 1$ and the scheme is unconditionally unstable.

4.4 Analytic solutions of partial difference equations

As we can easily observe from the numerical schemes investigated in section 4.3, each numerical scheme constitutes a partial difference equation. In this section we will investigate each numerical scheme from the point of view of a partial difference equation. For the solution we will use a basic technique which can be found in several classic books. This technique make use of the operators defined as

$$\begin{aligned}
E_1 u_i^j \equiv E_1 u = u_{i+1}^j, \quad E_2 u_i^j \equiv E_2 u = u_i^{j+1}, \\
E_{-1} u_i^j \equiv E_{-1} u = u_{i-1}^j, \quad E_{-2} u_i^j \equiv E_{-2} u = u_i^{j-1}
\end{aligned} \tag{4.4.1}$$

and the fact that these operator act on different variables and do not interfere with each other.

FTBS

We begin with the FTBS scheme:

$$u_i^{j+1} = u_i^j - 2\frac{\Delta t}{\Delta x}\left(u_i^j - u_{i-1}^j\right) \Leftrightarrow u_i^{j+1} = u_i^j - 2\lambda u_i^j + 2\lambda u_{i-1}^j \tag{4.4.2}$$

where $\lambda = \Delta t/\Delta x$.

The initial condition accompanying the pde (4.2.1) is $u(x,0) = x$ which gives rise to the following condition for (4.4.2):

$$u(i,1) = (i-1)\Delta x.$$

Equation (4.4.2) is a partial difference equation of second order with respect to i and second order with respect to j. It is worth mentioning that the pde (4.2.1) is of first order with respect to both x or t.

Thus, using the operators defined in (4.4.1), equation (4.4.2) take the form

$$E_2 u = u - 2\lambda u + 2\lambda E_{-1} u \Leftrightarrow E_2 u = (1 - 2\lambda) u + 2\lambda E_{-1} u$$

$$u = (1 - 2\lambda + 2\lambda E_{-1})^{j-1} a(i)$$

where $a(i)$ is an arbitrary sequence. However, from the initial condition, we obtain

$$u(i, 1) = (i - 1) \Delta x \Leftrightarrow a(i) = (i - 1) \Delta x$$

and thus

$$u(i, j) = (1 - 2\lambda + 2\lambda E_{-1})^{j-1} (i - 1) \Delta x$$

For $\lambda \neq 1/2$ we have:

$$u(i, j) = (1 - 2\lambda)^{j-1} \left(1 + \frac{2\lambda}{1 - 2\lambda} E_{-1} \right)^{j-1} (i - 1) \Delta x \Longleftrightarrow$$

$$u(i, j) = (1 - 2\lambda)^{j-1} \sum_{m=0}^{j-1} \left(\frac{2\lambda}{1 - 2\lambda} \right)^m E_{-1}^m (i - 1) \Delta x \Longleftrightarrow$$

$$u(i, j) = (1 - 2\lambda)^{j-1} \Delta x \sum_{m=0}^{j-1} \binom{j-1}{m} \frac{2^m \lambda^m}{(1 - 2\lambda)^m} (i - m - 1) \Longleftrightarrow$$

$$u(i, j) = (i - 1) \Delta x - 2 (j - 1) \Delta t \tag{4.4.3}$$

For

$$\lambda = \frac{1}{2} \Rightarrow \frac{\Delta t}{\Delta x} = \frac{1}{2} \Rightarrow \Delta x = 2\Delta t$$

it is obtained that

$$u(i, j) = (1 - 1 + E_{-1})^{j-1} (i - 1) \Delta x = (i - j) \Delta x \tag{4.4.4}$$

We observe that both (4.4.3) and (4.4.4) have in reality the same type and can be described by the relation

$$u(i, j) = (i - 1) \Delta x - 2 (j - 1) \Delta t, \quad for \quad \lambda > 0 \tag{4.4.5}$$

The obtained solution (4.4.5) converges to the analytic solution of the original partial differential equation (4.2.1) since

$$\lim_{\Delta x \to 0} \lim_{\Delta t \to 0} u(i, j) = \lim_{\Delta x \to 0} \lim_{\Delta t \to 0} \left[\left(\frac{x}{\Delta x} - 1 \right) \Delta x - 2 \left(\frac{t}{\Delta t} - 1 \right) \Delta t \right] = x - 2t$$

FTCS

Using the FTCS scheme it is found that the resulting difference equation is

$$u_i^{j+1} = u_i^j - \lambda \left(u_{i+1}^j - u_{i-1}^j \right) \Rightarrow u_i^{j+1} = u_i^j - \lambda u_{i+1}^j + \lambda u_{i-1}^j$$

where $\lambda = \Delta t / \Delta x$. This difference equation is of first order with respect to j and second order with respect to i.

By using the classical operators, as before, we have

$$E_2 u = u - \lambda E_1 u + \lambda E_{-1} u \Leftrightarrow E_2 u = \left(1 - \lambda E_1 + \lambda E_{-1} \right) u$$

from which it follows

$$u = \left(1 - \lambda E_1 + \lambda E_{-1} \right)^{j-1} a(i),$$

where $a(i)$ is an arbitrary sequence. From the corresponding initial condition we have

$$u(i,1) = (i-1)\,\Delta x \Leftrightarrow a(i) = (i-1)\,\Delta x$$

Thus, the solution takes the form

$$u(i,j) = \left[1 + \lambda \left(E_{-1} - E_1 \right) \right]^{j-1} (i-1)\,\Delta x \Longleftrightarrow$$

$$u(i,j) = \sum_{m=0}^{j-1} \binom{j-1}{m} \lambda^m \left(E_{-1} - E_1 \right)^m (i-1)\,\Delta x \Longleftrightarrow$$

$$u(i,j) = \sum_{m=0}^{j-1} \binom{j-1}{m} \lambda^m \sum_{n=0}^{m} \binom{m}{n} E_{-1}^{m-n} (-1)^n E_1^n (i-1)\,\Delta x \Longleftrightarrow$$

$$u(i,j) = \Delta x \sum_{m=0}^{j-1} \binom{j-1}{m} \lambda^m \sum_{n=0}^{m} \binom{m}{n} (-1)^n (i-1-m+2n)$$

or after some manipulation

$$u(i,j) = (i-1)\,\Delta x - 2\,(j-1)\,\Delta t \tag{4.4.6}$$

which is the same with that calculated using the previous FTBS scheme.

FTFS

For the FTFS we have

$$u_i^{j+1} = u_i^j - 2\lambda \left(u_{i+1}^j - u_i^j \right) = u_i^j - 2\lambda u_{i+1}^j + 2\lambda u_i^j \tag{4.4.7}$$

This difference equation is of first order with respect to j and first order with respect to i.

Using the classical shift operators we have

$$E_2 u = (1 + 2\lambda)\, u - 2\lambda E_1 u \iff$$

$$u\,(i,j) = (1 + 2\lambda - 2\lambda E_1)^{j-1}\, a(i)$$

The initial condition $u(x,0) = x$ gives for the corresponding discrete form that

$$j = 1: \quad u(i,1) = (i-1)\Delta x \Rightarrow a(i) = (i-1)\,\Delta x$$

Hence, we have that

$$u\,(i,j) = (1 + 2\lambda - 2\lambda E_1)^{j-1}\,(i-1)\,\Delta x$$

For $\lambda \neq -\dfrac{1}{2}$ we have

$$u\,(i,j) = (1 + 2\lambda)^{j-1}\left[1 - \frac{2\lambda}{1+2\lambda}\right]^{j-1}(i-1)\,\Delta x \iff$$

$$u(i,j) = (1 + 2\lambda)^{j-1}\sum_{m=0}^{j-1}\binom{j-1}{m}(-1)^m \left(\frac{2\lambda}{1+2\lambda}E_1\right)^m (i-1)\,\Delta x \iff$$

$$u(i,j) = (1 + 2\lambda)^{j-1}\sum_{m=0}^{j-1}\binom{j-1}{m}(-1)^m \frac{2^m\lambda^m}{(1+2\lambda)^m}(i+m-1)\,\Delta x \iff$$

$$u(i,j) = -\Delta x\,[1 - i + 2\,(-1 + j)\,\lambda] = (i-1)\,\Delta x - 2\,(j-1)\,\Delta t \qquad (4.4.8)$$

For $\lambda = -\frac{1}{2} \Rightarrow \frac{\Delta t}{\Delta x} = -\frac{1}{2} \Rightarrow \Delta x = -2\Delta t$ we also have

$$u(i,j) = (1 - 1 + E_1)^{j-1}\,(i-1)\,\Delta x = E_1^{j-1}\,(i-1)\,\Delta x = (i+j)\,\Delta x \qquad (4.4.9)$$

The two types of solution (4.4.8) and (4.4.9) can be combined in the following one relation for all λ

$$u(i,j) = (i-1)\,\Delta x - 2\,(j-1)\,\Delta t$$

Again, we observe that the above type of the discrete solution is the same as the type we obtained using the previous schemes.

Leap–Frog

By using the Leap–Frog scheme with forward differences for space, we have

$$u_i^{j+1} = u_i^{j-1} - 4\lambda \left(u_{i+1}^j - u_i^j \right) \tag{4.4.10}$$

which is a difference equation of second order with respect to j and first order with respect to i. We have

$$E_2 u = E_{-2} u - 4\lambda \left(E_1 u - u \right) \Leftrightarrow 4\lambda \left(E_1 u - u \right) = E_{-2} u - E_2 u \overset{\lambda \neq 0}{\Leftrightarrow}$$

$$E_1 u - u = \frac{1}{4\lambda} \left(E_{-2} u - E_2 u \right) \Leftrightarrow E_1 u = \left[1 + \frac{1}{4\lambda} \left(E_{-2} - E_2 \right) \right] u \Longleftrightarrow$$

$$u(i,j) = \left[1 + \frac{1}{4\lambda} \left(E_{-2} - E_2 \right) \right]^{i-1} a(j)$$

At this point, it is significantly difficult to proceed with the solution using the known initial condition $u(i,1) = (i-1)\Delta x$, since the partial difference equation is of second order with respect to j. So, in order to proceed and find the solution in closed form, we use the known analytic solution $u(x,t) = x - 2t$ to construct a condition towards the x direction. Thus, for this case only, we use as condition that $u(0,t) = -2t$, which in discrete mode is made

$$i = 1: \quad u(1,j) = a(j) \Leftrightarrow a(j) = -2(j-1)\Delta t \tag{4.4.11}$$

By using the above condition the solution takes the form

$$u(i,j) = \left[1 + \frac{1}{4\lambda} \left(E_{-2} - E_2 \right) \right]^{i-1} (-2(j-1)\Delta t) \Longleftrightarrow$$

$$u(i,j) = \sum_{m=0}^{i-1} \binom{i-1}{m} \frac{1}{4^m \lambda^m} \left(E_{-2} - E_2 \right)^m (-2(j-1)\Delta t) \Longleftrightarrow$$

$$u(i,j) = \sum_{m=0}^{i-1} \binom{i-1}{m} \frac{1}{4^m \lambda^m} \sum_{n=0}^{m} \binom{m}{n} E_{-2}^{m-n} (-1)^n E_2^n (-2(j-1)\Delta t) \Longleftrightarrow$$

$$u(i,j) = -2\Delta t \sum_{m=0}^{i-1} \binom{i-1}{m} \frac{1}{4^m \lambda^m} \sum_{n=0}^{m} \binom{m}{n} (-1)^n (j + 2n - m - 1)$$

and after some manipulation we conclude that

$$u(i,j) = \Delta t \left(\frac{i-1}{\lambda} - 2(j-1) \right) = (i-1)\Delta x - 2(j-1)\Delta t$$

which converges to the analytic solution of the original partial equation (4.2.1) and is the same with the type obtained using all the previous schemes.

4.5   Numerical results.

We perform now some numerical computations for equation (4.2.1). The discretization schemes we will use, are the discretization schemes leading to the corresponding difference equations solved analytically in session 4.4.

At the same figures we plot:

- The exact solution of the original partial differential equation (4.2.1), which is $u(x,t) = x - 2t$, as a straight line.

- The analytical solution of the corresponding, to each discretization scheme, difference equation. These solutions were derived in section 4.4 and turned out to be $u(i,j) = (i-1)\,\Delta x - 2\,(j-1)\,\Delta t$ for all the discretization schemes used. This solution, for each scheme, is marked in the figures as "Difference" and the corresponding points are represented by cross.

- The computationally estimated solution using the discretization scheme leading to the same difference equation. The points of the solution are plotted as hollow circles.

Figure 4.2 shows the results for the FTBS discretization scheme for $\Delta t = 0.01$ and $\Delta x = 0.02$. It is reminded that by the Von Neumann stability analysis all the discretization schemes are proved to be unconditionally unstable except the FTBS scheme which is stable for $\lambda \leq 1/2 \Rightarrow 2\Delta t \leq \Delta x$. Thus, the values of time and space step satisfy the stability condition for this scheme. The $u(x,t)$ function is pictured with x at different times ($t =$0, 0.2, 0.4, 0.6, 0.8 and 1.0). The numerical results are in accordance with the exact solution and the solution estimated analytically by the corresponding difference equation. As we expected, the computed solution remains stable as time passes.

Results for the FTBS scheme are shown in Figure 4.3 for $\Delta t = 0.01$ and $\Delta x = 0.0125$ which do not satisfy the Von Neumann stability condition. It is apparent that the computational solution remains stable for small times. For $t = 0.4$ the computationally estimated solution is stable whereas, for time $t = 0.6$ is unstable and the computed points are blasted far away from the true solution. It is observed that the computational solution start to blast for smaller x as t passes. This is normal, since the scheme is unstable and the errors increase from one iteration of time to another.

Figure 4.4 shows results for the FTCS scheme for $\Delta t = 0.01$ and $\Delta x = 0.02$. It is remarkable that the computational solution is in accordance with the exact and the analytic solution of the corresponding difference equation for all times plotted. In fact the behaviour of the computational solution for this scheme is very similar with that of the FTBS although this scheme is unstable according to the stability analysis.

Figure 4.5 shows the solution derived for $\Delta t = 0.01$ and $\Delta x = 0.0125$, which are the same steps with FTBS in Figure 4.3. The computational solution here has a behaviour similar to the solution computed using the FTBS scheme and after short times of stable solutions at $t = 0.6$ starts oscillations and hereafter blasts away.

We perform now additional computations for the FTCS scheme which, until now, exhibits very similar behaviour with the FTBS scheme. Figure 4.6 shows the derived solutions for times $t \geq 1.0$. We observe that after a short period of time beyond $t = 1.0$ the computational solution starts to oscillate (at $t = 1.1$) and it is blasted for times greater than $t = 1.2$. It is noted, that computations were performed for the stable FTBS as well as for greater times. It is observed that the computed solution for the FTBS remains stable for very large times as $t = 100$ or $t = 1000$.

It is also observed that the computational solution for the schemes FTBS and FTCS is stable for times up to $t = 1$. The derivation of a stable computational solution for the FTCS, although this scheme is proved to be unstable, is not in contradiction with this theoretical result. The results of the Von Neumann stability analysis are valid as $t \to +\infty$. So, if a scheme is proved unstable, it is possible to evaluate numerically the correct approximate solution for some finite time steps but it is also sure that with the time propagation instabilities will appear sooner or later.

This phenomenon of evaluating the correct approximate computational solution for a finite time step, using an unstable numerical scheme, could be justified because we use consistent schemes (the solution of the difference equation tends to the true solution of the original differential equation as $\Delta t, \Delta x \to 0$) and the accumulation errors for some finite times are increasing, but not so significantly as to cause loss of the correct approximate solution.

Figure 4.7 shows results for $\Delta t = 0.01$ and $\Delta x = 0.08$. With these time and space steps we attain to derive a stable computational solution until $t = 1.0$. The computational solution for the same scheme is made sooner unstable, than the other schemes, for the steps $\Delta t = 0.01$ and $\Delta x = 0.02$ and it has already blasted from the correct solution at time $t = 0.4$ as shown in Figure 4.8. We perform some additional computations using the steps $\Delta t = 0.01$ and $\Delta x = 0.06$ shown in Figure 4.9. We observe that this time the instabilities are introduced to the solution at $t = 0.8$ and at $t = 1.0$ the solution has already blasted.

From figures 4.7, 4.8 and 4.9 we observe that the rate of transition to instability of the computational solution depends on the relation of the time and space steps. With fixed $\Delta t = 0.01$ more stable solutions are produced for the greater $\Delta x = 0.08$. As Δx decreases, (with Δt fixed) instabilities are increasing more rapidly with the increase of time t and this is a verification of a known empirical rule.

The above mentioned behaviour is observed for the Leap–Frog with Forward Space (LFFS) discretization scheme. We observe that this numerical scheme is "more" unstable than the other numerical schemes in the sense that Δx should increased up to 0.1 in order to attain stable solution up to $t = 1.0$. Figures 4.10 and 4.11 show this behaviour.

It is remarkable that the solution estimated by the analytical solution of the corresponding difference equation for all schemes, is always stable and follows the analytical solution irrespectively the relation of the steps Δx and Δt. If we notice the analytic discrete solution we derive from all the corresponding difference equations (to the discretization schemes) we used, it's type is

$$u(i,j) = (i-1)\Delta x - 2(j-1)\Delta t, \quad i = 1, \cdots M \quad j = 1, \cdots, N$$

At this point we stress out that this solution is attained from different difference equations. Clearly, the use of different discretization schemes leads to difference equations different with each other and, in many cases, with different orders with respect i or j.

The analytic solution of the original partial differential equation (4.2.1) is

$$u(x,t) = x - 2t$$

Thus, the above discrete solution is nothing less that the discrete form of the analytic solution in a grid pictured in Figure 4.1 since $(i-1)\Delta x \to x$ and $(j-1)\Delta t \to t$. The discrete solution is estimated directly from the type $u(i,j) = (i-1)\Delta x - 2(j-1)\Delta t$ and not through a computational marching procedure as it happens by estimating the computational solution and for this reason there is no matter of accumulation errors and need for Von Neumann stability analysis. Moreover, there is no other restriction arising during the analytic solution of the corresponding difference equations and the estimation of the discrete solution for these simple cases is quite straightforward.

Summarizing, from the original partial differential equation we derive, by the application of a numerical scheme, a partial difference equation consistent to the original differential equation. If we solve analytically the partial difference equation and no limitations or restrictions arise from this procedure, we do not need any further stability analysis for the computation of the discrete solution. What we only need, is to use such time and space steps, as to hold the consistency of the difference to differential equation.

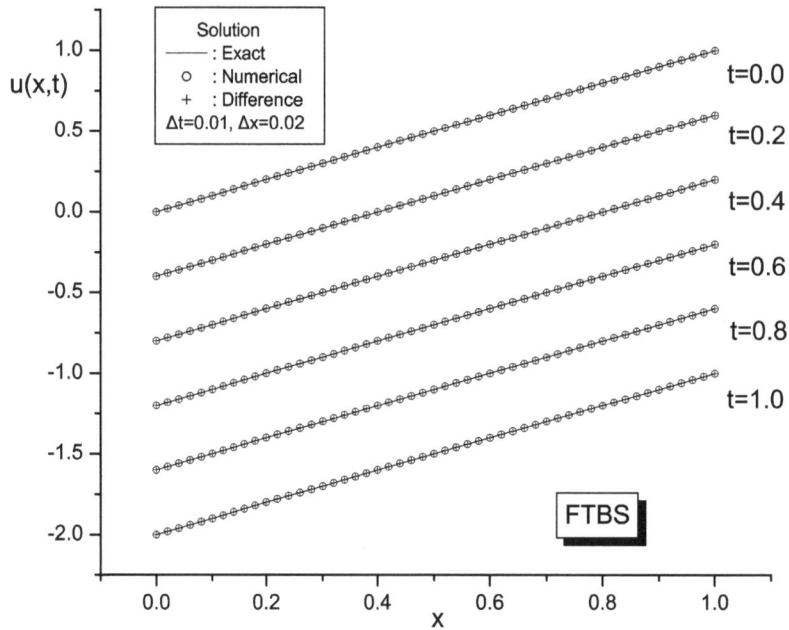

Figure 4.2: Results for FTBS scheme, for $\Delta x = 0.02$ and $\Delta t = 0.01$

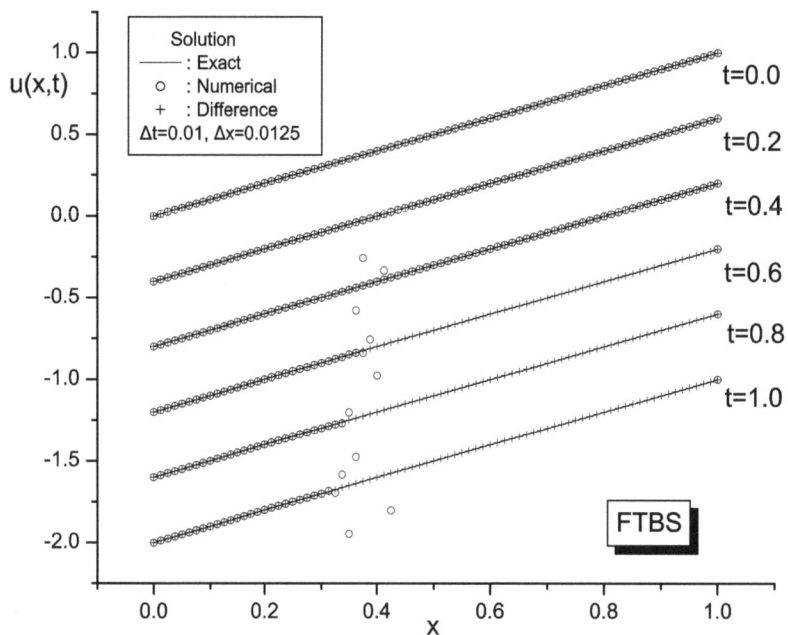

Figure 4.3: Results for FTBS scheme, for $\Delta x = 0.0125$ and $\Delta t = 0.01$

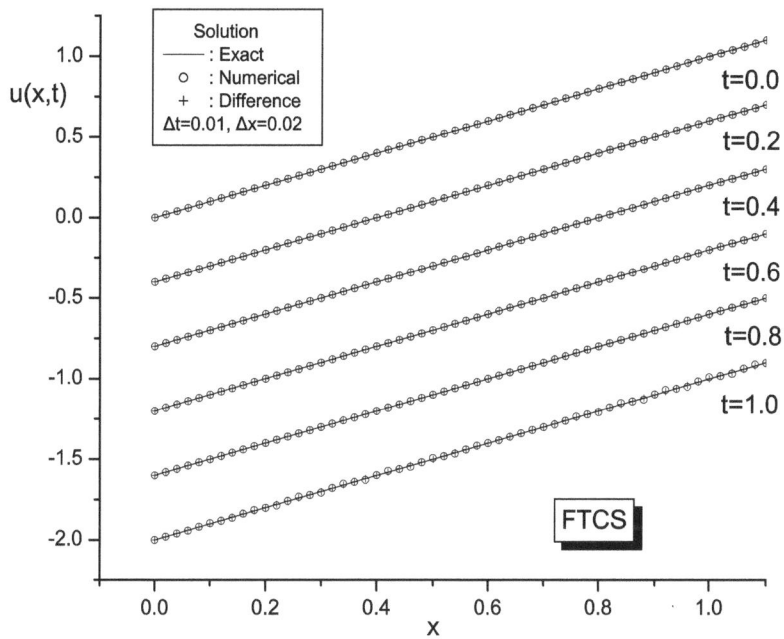

Figure 4.4: Results for FTCS scheme, for $\Delta x = 0.02$ and $\Delta t = 0.01$

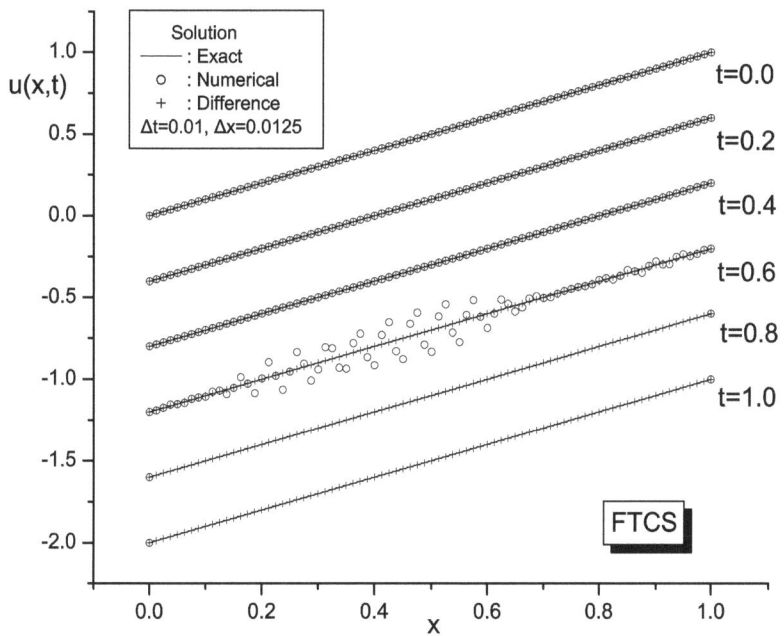

Figure 4.5: Results for FTCS scheme, for $\Delta x = 0.0125$ and $\Delta t = 0.01$

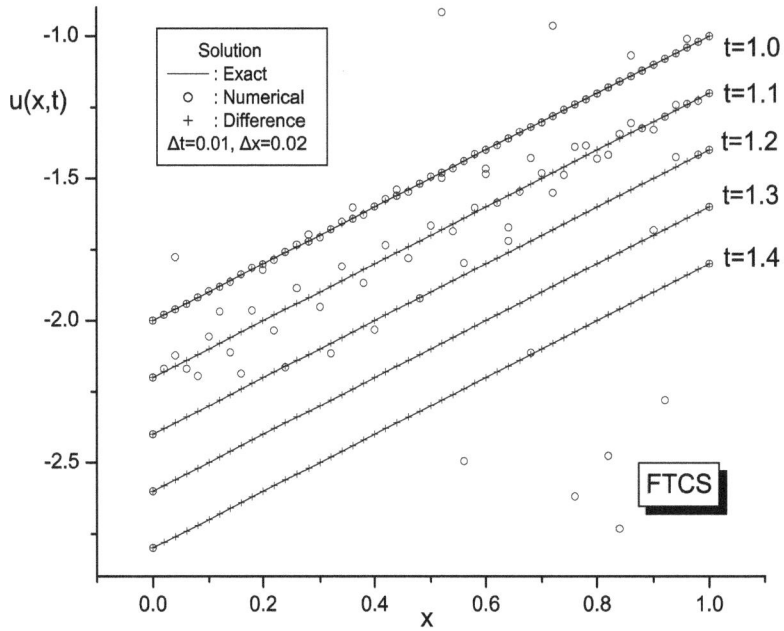

Figure 4.6: Results for FTCS scheme, for $\Delta x = 0.02$ and $\Delta t = 0.01$

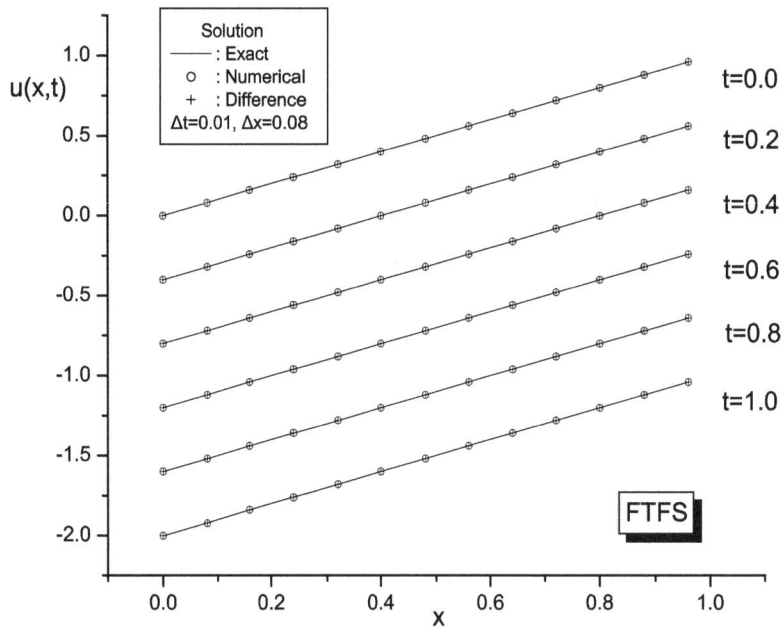

Figure 4.7: Results for FTFS scheme, for $\Delta x = 0.08$ and $\Delta t = 0.01$

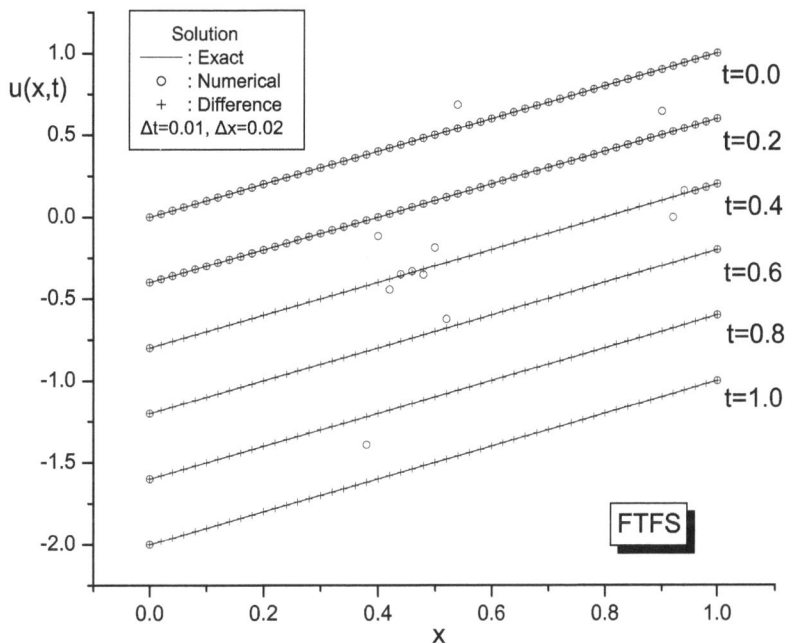

Figure 4.8: Results for FTFS scheme, for $\Delta x = 0.02$ and $\Delta t = 0.01$

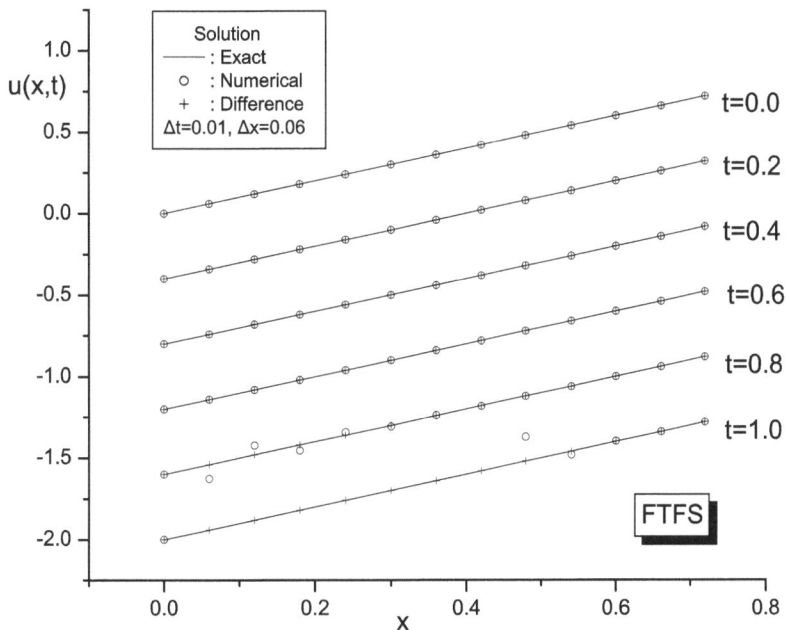

Figure 4.9: Results for FTFS scheme, for $\Delta x = 0.06$ and $\Delta t = 0.01$

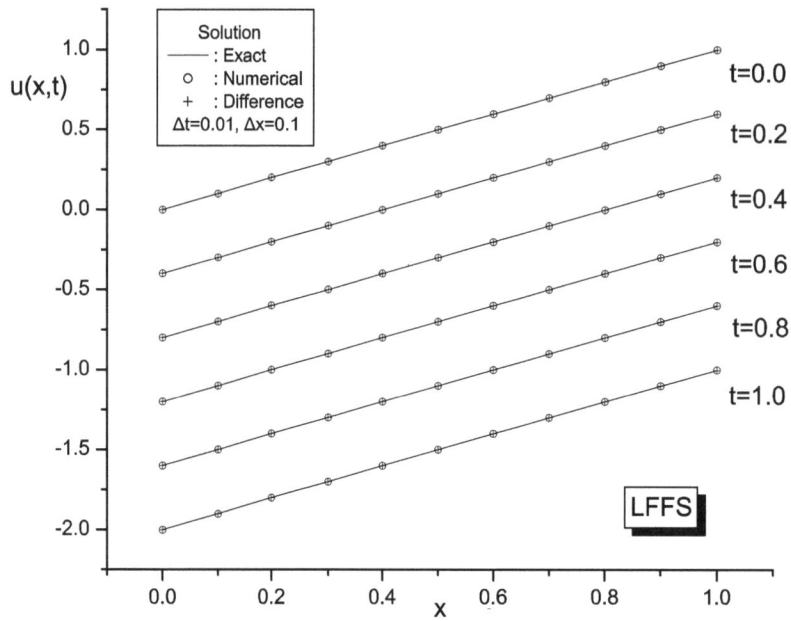

Figure 4.10: Results for LFFS scheme, for $\Delta x = 0.1$ and $\Delta t = 0.01$

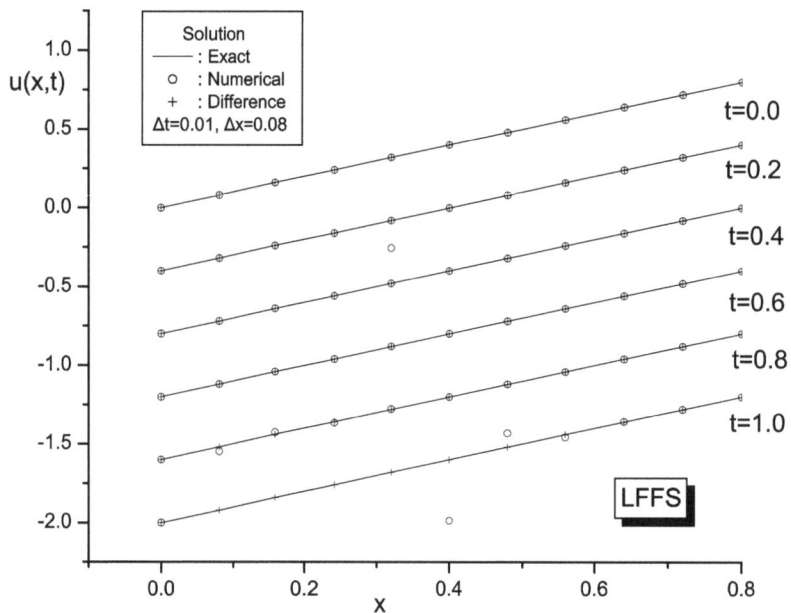

Figure 4.11: Results for LFFS scheme, for $\Delta x = 0.08$ and $\Delta t = 0.01$

4.6   Conclusions

We studied the connection of the difference equations and numerical schemes. We used the well known finite differences scheme to dicretize the simple linear pde $u_t + 2u_x = 0$ with initial condition $u(x,0) = x$. By using different discretization schemes we derived different, consistent to the original pde, numerical schemes which constitute corresponding difference equations. We solved each one of the difference equations analytically and we arrived for all, to the same discrete type solution. Is is noted that the difference equations, arising after applying different discretization schemes, are not the same with each other, and in many cases are of different order with respect to the time or space variable.

We also used the above mentioned discretizations (and the corresponding difference equations) for the numerical solution of the above mentioned pde. This time, for each difference equation, the solution is computed using a marching procedure. The Von Neumann stability analysis of each numerical scheme results that all the schemes, except one, are numerically unstable.

Results of the numerical simulations show that it is possible to achieve correct approximate solution, for finite times, using unstable numerical schemes. For unstable schemes the time point, beyond of which instabilities begin to occur and the solution is lost, depends on the relation of the time and space steps. On the other hand, the solution estimated by the analytical solution of the corresponding (to the numerical scheme) partial difference equation is always stable. In this case there is no issue of stability since the discrete solution is calculated straightforward from it's type, without using iterative (or marching) procedure. The only limitation for the time and space step is to use such steps so as to have consistency of the difference equation to the original continuum pde $u_t + 2u_x = 0$.

Thus, it is emerged that the analytic solution of difference equations, resulting from application of numerical schemes, could be of extremely importance for the estimation of the solution of a pde. The analytic solution of a partial difference equation is not an easy task but whatever information could be extracted from it, could be very useful for the estimation of a solution. For example, one can have asymptotic analysis results or bounds for the solution. Such results, which are not affected by possible instabilities of the numerical scheme during an iterative procedure, could be very useful for the estimation or verification of the numerical findings.

References

[1] R. P. Agarwal. *Difference equations and inequalities. Theory, methods and applications.* Marcel Dekker, 1992.

[2] W. F. Ames. *Numerical Methods of Partial Differential Equations.* Academic Press, 1977.

[3] J.D.Jr. Anderson. *Computational Fluid Dynamics.* McGraw–Hill, Inc, New York.

[4] J. P. Boyd. *Chebyshev and Fourier Spectral Methods*. Dover Publications, INC, Mineola, New York, 2001.

[5] C. Canuto, M.Y. Hussaini, A. Quarteroni, and T. A. Zang. *Spectral Methods in Fluid Dynamics*. Springer–Verlag, New York.

[6] T. J. Chung. *Computational Fluid Dynamics*. Cambridge University Press, 2002.

[7] C.A.J. Fletcher. *Computational Techniques for Fluid Dynamics*. Springer–Verlag, New York.

[8] P. D. Lax. Numerical solution of partial differential equations. *Amer. Math. Monthly*, 72(2):74–84, 1965.

[9] R. E. Mickens. *Difference equations. Theory and applications*. Van Nostrand Reinhold Co., 1990.

[10] R. E. Mickens. *Advances in the Applications of Non–Standard Finite Difference Schemes*. World Scientific Publishing Co Pte Ltd, 2005.

[11] G. C. O Brien, M. A. Hyman, and S. Kaplan. A study of the numerical solution partial differential equation. *J. Math. Physics*, 29:223–251, 1950.

[12] R. D Richtmyer and K. W. Morton. *Difference Methods for initial value problems*. Wiley Interscience, 1967.

[13] G. D. Smith. *Numerical solution of partial differential equation, Finite Difference Method*. Oxford University Press, 1978.

[14] H. K. Versteeg and W. Malalasekera. *An introduction to computational fluid dynamics, The Finite Volume Method*. Longman Scientific and Technical, 1995.

Index